# NEUROTECHNOLOGY

## BRAIN-COMPUTER-INTERFACE AND THE FUTURE OF HUMANITY

### DAVID SANDUA

Neurotechnology.
© David Sandua 2024. All rights reserved.
Electronic and paperback edition.

*"The integration of technology and the human mind will be the biggest revolution since the discovery of fire."*

*Marvin Minsky*
*(Computer scientist and co-founder of the MIT AI Lab)*

# INDEX

**I. INTRODUCTION** .................................................................................................... 11
- DEFINITION OF NEUROTECHNOLOGY ................................................................... 12
- SIGNIFICANCE OF BCI ............................................................................................ 13
- THESIS STATEMENT ............................................................................................... 15

**II. HISTORICAL DEVELOPMENT OF NEUROTECHNOLOGY** ................................... 17
- EARLY STUDIES ON BRAIN FUNCTION ................................................................. 18
- EVOLUTION OF BCI TECHNOLOGY ....................................................................... 19
- MILESTONES IN NEUROTECHNOLOGY RESEARCH .............................................. 20

**III. TYPES OF BCIS** .................................................................................................. 22
- INVASIVE BCIS ....................................................................................................... 22
- NON-INVASIVE BCIS .............................................................................................. 24
- HYBRID BCIS .......................................................................................................... 25

**IV. APPLICATIONS OF NEUROTECHNOLOGY IN HEALTHCARE** ............................ 27
- NEUROREHABILITATION ........................................................................................ 28
- NEUROPROSTHETICS ............................................................................................ 29
- NEURAL IMAGING AND DIAGNOSIS ..................................................................... 31

**V. ETHICAL CONSIDERATIONS IN NEUROTECHNOLOGY** ..................................... 33
- PRIVACY AND DATA SECURITY ............................................................................. 34
- INFORMED CONSENT AND AUTONOMY ............................................................... 35
- EQUITY AND ACCESS TO NEUROTECHNOLOGICAL ADVANCEMENTS ................ 36

**VI. NEUROTECHNOLOGY IN COGNITIVE ENHANCEMENT** ................................... 38
- MEMORY AUGMENTATION .................................................................................... 39
- COGNITIVE SKILLS TRAINING ............................................................................... 40
- ETHICAL IMPLICATIONS OF COGNITIVE ENHANCEMENT .................................... 41

**VII. NEUROTECHNOLOGY AND HUMAN-MACHINE INTERACTION** ..................... 43
- AUGMENTED REALITY AND VIRTUAL REALITY ..................................................... 44
- NEUROFEEDBACK IN GAMING .............................................................................. 45
- IMPLICATIONS FOR EDUCATION AND TRAINING ................................................ 46

**VIII. NEUROTECHNOLOGY AND NEUROETHICS** ................................................... 48
- NEURODIVERSITY AND INCLUSIVITY .................................................................... 49
- NEUROEXISTENTIALISM AND IDENTITY ............................................................... 50
- REGULATION AND GOVERNANCE OF NEUROTECHNOLOGIES ........................... 51

**IX. CHALLENGES AND FUTURE DIRECTIONS IN NEUROTECHNOLOGY** .............. 53
- TECHNOLOGICAL LIMITATIONS AND INNOVATIONS ........................................... 54
- ETHICAL AND SOCIAL IMPLICATIONS .................................................................. 55
- INTERDISCIPLINARY COLLABORATION IN ADVANCING NEUROTECHNOLOGY .... 56

**X. NEUROTECHNOLOGY AND NEUROAESTHETICS** .............................................. 58
- IMPACT OF NEUROTECHNOLOGY ON ARTISTIC EXPRESSION ............................ 59
- NEUROAESTHETIC EXPERIENCES IN VIRTUAL REALITY ...................................... 60
- NEUROAESTHETICS IN DESIGN AND ARCHITECTURE ......................................... 61

**XI. NEUROTECHNOLOGY IN MILITARY AND DEFENSE** ........................................ 63
- BCIS FOR ENHANCED SOLDIER PERFORMANCE .................................................. 64
- NEUROENHANCEMENT IN MILITARY TRAINING .................................................. 65
- ETHICAL CONSIDERATIONS IN WEAPONIZED NEUROTECHNOLOGY ................. 66

## XII. NEUROTECHNOLOGY AND NEUROPLASTICITY .... 68
### Harnessing Neuroplasticity for Cognitive Enhancement .... 69
### Neurofeedback Training for Brain Plasticity .... 70
### Implications for Learning and Skill Acquisition .... 71

## XIII. NEUROTECHNOLOGY IN NEUROSCIENCE RESEARCH .... 73
### Advancements in Brain Mapping and Connectivity Studies .... 74
### Neuroimaging Techniques for Understanding Brain Function .... 75
### Neurotechnology in Studying Neurological Disorders .... 76

## XIV. NEUROTECHNOLOGY AND HUMAN RIGHTS .... 78
### Access to Neurotechnological Advancements as a Human Right .... 79
### Ethical Considerations in Cognitive Liberty .... 80
### Neurotechnology and the Right to Mental Privacy .... 81

## XV. NEUROTECHNOLOGY IN SPORTS PERFORMANCE .... 83
### BCIs for Athlete Training and Monitoring .... 84
### Cognitive Enhancement in Sports Psychology .... 85
### Ethical Implications of Performance-Enhancing Neurotechnology .... 86

## XVI. NEUROTECHNOLOGY AND ENVIRONMENTAL SUSTAINABILITY .... 89
### Applications of BCI in Environmental Monitoring .... 90
### Neurofeedback for Sustainable Behavior Change .... 91
### Neurotechnology in Promoting Eco-Friendly Practices .... 92

## XVII. NEUROTECHNOLOGY AND AGING POPULATION .... 94
### Cognitive Support for Elderly Individuals .... 95
### Neurorehabilitation for Age-Related Cognitive Decline .... 96
### Ethical Considerations in Enhancing Quality of Life for Seniors .... 97

## XVIII. NEUROTECHNOLOGY AND GLOBAL HEALTH .... 99
### BCI Applications in Developing Countries .... 100
### Neurotechnology for Mental Health Support .... 101
### Ethical Challenges in Implementing Neurotechnological Solutions .... 102

## XIX. NEUROTECHNOLOGY AND NEURODIVERSITY .... 104
### Enhancing Accessibility for Neurodiverse Individuals .... 105
### Supporting Neurodivergent Communities with BCI Technology .... 106
### Ethical Considerations in Neurotechnological Inclusivity .... 107

## XX. NEUROTECHNOLOGY AND NEUROETHICS IN EDUCATION .... 109
### Implementing BCI in Educational Settings .... 110
### Enhancing Learning and Cognitive Development with Neurotechnology .... 111
### Ethical Implications of Neuroenhancement in Education .... 112

## XXI. NEUROTECHNOLOGY AND EMOTIONAL INTELLIGENCE .... 114
### BCI Applications in Emotion Recognition .... 115
### Enhancing Emotional Regulation through Neurofeedback .... 116
### Ethical Considerations in Emotional Manipulation with Neurotechnology .... 117

## XXII. NEUROTECHNOLOGY IN LAW ENFORCEMENT AND CRIMINAL JUSTICE .... 119
### BCI for Lie Detection and Interrogation .... 120
### Neuroimaging in Criminal Profiling and Evidence Analysis .... 121
### Ethical Challenges in the Use of Neurotechnology in Legal Contexts .... 122

## XXIII. NEUROTECHNOLOGY AND WORKPLACE PRODUCTIVITY .... 124
### BCI for Enhancing Focus and Task Performance .... 125
### Neurofeedback Training for Stress Management in the Workplace .... 126
### Ethical Considerations in Employee Monitoring with Neurotechnology .... 127

## XXIV. NEUROTECHNOLOGY AND PERSONALIZED MEDICINE ............129
- PRECISION MEDICINE APPROACHES WITH BCI DATA.................................. 130
- TAILORING TREATMENT PLANS USING NEUROTECHNOLOGICAL INSIGHTS.................. 131
- ETHICAL IMPLICATIONS OF PERSONALIZED HEALTHCARE THROUGH NEUROTECHNOLOGY ........ 132

## XXV. NEUROTECHNOLOGY AND SOCIAL IMPACT ..............................134
- ADDRESSING SOCIAL INEQUALITY THROUGH ACCESS TO NEUROTECHNOLOGICAL ADVANCEMENTS...... 135
- PROMOTING EMPATHY AND UNDERSTANDING WITH BCI TECHNOLOGY........................... 136
- ETHICAL CONSIDERATIONS IN SHAPING SOCIAL NORMS THROUGH NEUROTECHNOLOGY ........... 137

## XXVI. NEUROTECHNOLOGY AND NEUROPHILOSOPHY............................139
- EXPLORING CONSCIOUSNESS AND IDENTITY WITH BCI ..................................... 140
- PHILOSOPHICAL IMPLICATIONS OF DIRECT BRAIN MANIPULATION ........................... 141
- ETHICAL CONSIDERATIONS IN NEUROPHILOSOPHICAL INQUIRY .............................. 142

## XXVII. NEUROTECHNOLOGY IN SPACE EXPLORATION ..........................144
- BCI FOR ASTRONAUT TRAINING AND MONITORING ........................................ 145
- ENHANCING COGNITIVE PERFORMANCE IN EXTREME ENVIRONMENTS ........................... 146
- ETHICAL CONSIDERATIONS IN SPACE TRAVEL WITH NEUROTECHNOLOGICAL SUPPORT............. 147

## XXVIII. NEUROTECHNOLOGY AND ARTISTIC INNOVATION ......................149
- BRAINWAVE ART AND NEUROFEEDBACK IN CREATIVE PROCESSES ............................ 150
- ENHANCING ARTISTIC EXPRESSION THROUGH BCI TECHNOLOGY ............................. 151
- ETHICAL IMPLICATIONS OF NEUROAESTHETIC EXPERIENCES ............................... 152

## XXIX. NEUROTECHNOLOGY AND NEUROETHICS IN ARTIFICIAL INTELLIGENCE (AI) .....154
- ETHICAL CONSIDERATIONS IN AI INTEGRATION WITH BCI TECHNOLOGY...................... 155
- IMPLICATIONS OF AI ALGORITHMS ON NEURAL DATA PRIVACY .............................. 156
- ENSURING ETHICAL AI DEVELOPMENT AND IMPLEMENTATION................................ 157

## XXX. NEUROTECHNOLOGY AND NEUROSECURITY ..............................159
- CYBERSECURITY CHALLENGES IN BCIS ................................................ 160
- SAFEGUARDING NEURAL DATA FROM UNAUTHORIZED ACCESS ................................ 161
- DEVELOPING SECURE PROTOCOLS FOR NEUROTECHNOLOGICAL SYSTEMS........................ 162

## XXXI. NEUROTECHNOLOGY AND NEURODIVERSITY ADVOCACY ....................164
- PROMOTING INCLUSIVITY AND ACCESSIBILITY FOR NEURODIVERSE INDIVIDUALS ............. 165
- ADVOCATING FOR ETHICAL USE OF BCI TECHNOLOGY IN NEURODIVERGENT COMMUNITIES ....... 166
- ADDRESSING STIGMA AND DISCRIMINATION THROUGH NEURODIVERSITY AWARENESS ............ 167

## XXXII. CONCLUSION .................................................170
- SUMMARY OF KEY FINDINGS ON NEUROTECHNOLOGY AND BCIS .............................. 171
- REFLECTION ON THE TRANSFORMATIVE POTENTIAL AND ETHICAL CONSIDERATIONS............. 172
- CALL TO ACTION FOR CONTINUED RESEARCH AND ETHICAL REFLECTION IN ADVANCING NEUROTECHNOLOGICAL INNOVATIONS .......................................................................... 173

## BIBLIOGRAPHY .....................................................175

# I. INTRODUCTION

Neurotechnology, specifically Brain-Computer-Interface (BCI), represents a groundbreaking field that has the potential to revolutionize the way humans interact with technology. By bridging the gap between the brain and external devices, BCIs offer the possibility of controlling machines through mere thought, opening up a myriad of possibilities for individuals with physical disabilities or limitations. This technology has the capacity to enhance communication, improve motor function, and even treat neurological disorders. As we delve deeper into the realm of neurotechnology, it becomes increasingly apparent that the development and application of BCIs have immense implications for the future of humanity. The evolution of BCIs can be traced back to early research in neuroscience, where pioneering discoveries laid the foundation for today's advanced systems. From rudimentary experiments to sophisticated technologies, BCIs have come a long way in enabling direct communication between the brain and external devices. The basic principles of how BCIs operate involve the use of sensors to detect brain signals, signal processing to decode these signals, and algorithms to translate them into commands for external devices. Whether invasive or non-invasive, BCIs offer unique opportunities for individuals to interact with the world around them in unprecedented ways. In the medical field, BCIs have shown great promise in improving the quality of life for individuals with disabilities. From mind-controlled prostheses to neurorehabilitation for stroke patients, BCIs have opened up new avenues for treatment and intervention. The potential of BCIs to address neurological and psychiatric disorders showcases the transformative power of this technology. As we navigate the complexities of integrating BCIs into various domains, it is essential to consider the ethical and social challenges that come with direct manipulation of the human brain. As we move towards a future where BCIs are seamlessly integrated into everyday life, it is crucial to address these challenges to ensure the responsible and ethical advancement of neurotechnology for the benefit of society as a whole.

## Definition of Neurotechnology

Neurotechnology, at its core, refers to the application of electronics and engineering principles to the field of neuroscience. This interdisciplinary approach allows for the development of devices and technologies that can interact directly with the nervous system, particularly the brain. One of the most revolutionary forms of neurotechnology is the BCI, which enables communication and control between the brain and external devices without the need for traditional means such as speech or movement. BCIs can interpret brain signals and translate them into commands that can be used to operate computers, prosthetic limbs, or even control devices in smart homes. This seamless integration of human cognition and technology holds tremendous potential for transforming the way we interact with machines and the world around us. The history and evolution of BCIs can be traced back to the early explorations of neural communication and brain activity. Early research in neuroscience laid the foundation for understanding how the brain generates electrical signals that can be detected and utilized for communication purposes. The first experiments with BCIs involved invasive procedures where electrodes were implanted directly into the brain to record neural activity. Over time, advances in technology have led to the development of non-invasive BCIs that can capture brain signals through external sensors placed on the scalp. These advancements have made BCIs more accessible and practical for a wider range of applications, from medical rehabilitation to gaming and entertainment. In the realm of medical and therapeutic applications, BCIs have shown great promise in providing solutions for individuals with motor disabilities or neurological conditions. Mind-controlled prostheses powered by BCIs offer new possibilities for amputees to regain dexterity and independence in their daily lives. Similarly, stroke patients undergoing neurorehabilitation can benefit from the use of BCIs to retrain their neural pathways and restore motor function. The potential of BCIs to improve the quality of life for individuals with neurological and psychiatric disorders is a driving force behind ongoing research and development in the field of neurotechnology. As the technology continues to advance, the future of BCIs holds the promise of unlocking new ways for humans to interact with machines and improve the overall well-being of individuals facing physical and cognitive challenges.

## Significance of BCI

As BCIs continue to advance, their significance becomes increasingly apparent in various fields. One key area where BCIs are making a significant impact is in medicine and therapeutics. By utilizing BCIs in the treatment of motor disabilities, such as with mind-controlled prostheses, individuals who have lost the ability to control their limbs can regain a sense of independence and mobility. BCIs are being used in neurorehabilitation for stroke patients, helping to enhance recovery outcomes and improve quality of life for those affected by neurological conditions. The potential of BCIs to treat a wide range of neurological and psychiatric disorders holds promise for revolutionizing the way these conditions are managed and treated. In the realm of entertainment and communication, BCIs are opening up new possibilities for immersive experiences and enhanced communication. Mind-controlled video games and virtual reality experiences offer a glimpse into the future of entertainment, where users can interact with technology using only their thoughts. BCIs have the potential to improve communication for individuals with physical limitations, providing a means for expressing thoughts and emotions that may have been previously difficult or impossible to convey. The integration of BCIs in the entertainment industry not only offers exciting new opportunities for engagement but also highlights the ways in which technology can break down barriers and enhance human interaction. Despite the numerous benefits and advancements in BCIs, there are also ethical and social challenges that must be addressed. Privacy and security concerns related to the collection and usage of neural data raise important questions about consent and data protection. The potential risks of technology dependence and abuse, as well as the ethical implications of directly manipulating the human brain, underscore the importance of carefully considering the impact of BCIs on individuals and society as a whole. By recognizing and addressing these challenges, stakeholders can work towards ensuring that BCIs are developed and implemented in a responsible and ethical manner, ultimately maximizing their potential to improve lives and create positive societal change.

## Thesis Statement

The development and evolution of BCIs can be traced back to early research and fundamental discoveries in neuroscience. Scientists and researchers have long been intrigued by the possibility of connecting the human brain directly to machines, leading to the initial experiments and applications of BCI technology. From the rudimentary devices of the past to the advanced systems of today, the evolution of BCIs showcases the relentless pursuit of understanding the intricate workings of the brain and harnessing its potential for human-machine interaction. This journey through history highlights the innovative strides made in the field of neurotechnology, paving the way for a future where BCIs could shape the way we interface with technology. One of the key aspects of BCI technology is its operation, which involves a complex interplay of sensors, signal processing, and decoding algorithms. Understanding the basic principles behind how BCIs work is essential to grasp the transformative power they hold in various fields such as medicine, communication, and entertainment. By elucidating the differences between invasive and non-invasive BCIs, researchers can explore the multitude of applications that these technologies offer, from mind-controlled prostheses for individuals with motor disabilities to enhanced communication platforms for those with physical limitations. The intricate workings of BCI technology open up a realm of possibilities for improving the quality of life and expanding human capabilities. Looking towards the future, the emerging trends and technological advances in the field of BCIs herald a new era of human-machine symbiosis. As BCIs become more integrated into our daily lives, their impact on society and the way we interact with technology will undoubtedly be profound. By reflecting on the co-evolution of humans and technology, we can envision a world where the boundaries between the mind and machine blur, opening up new frontiers for exploration and discovery. The transformative potential of BCIs in enhancing the quality of life and unlocking new possibilities for humanity underscores the importance of continuing to address ethical and societal challenges in this rapidly evolving field.

## II. HISTORICAL DEVELOPMENT OF NEUROTECHNOLOGY

The historical development of neurotechnology can be traced back to early research and fundamental discoveries in neuroscience. In the late 1960s and 1970s, pioneering work by researchers such as Jacques Vidal laid the groundwork for the development of BCIs. Early experiments involved using electroencephalography (EEG) signals to control a cursor on a screen, demonstrating the potential for direct communication between the brain and external devices. These early developments marked the beginning of a transformative journey towards harnessing the power of neural signals for human-machine interaction. As research into BCIs progressed, the technology evolved from basic experimental setups to sophisticated systems capable of decoding complex brain signals. Today, BCIs encompass a range of components, including sensors to capture neural activity, signal processing algorithms to extract meaningful information, and decoding algorithms to translate neural signals into actionable commands. The distinction between invasive and non-invasive BCIs has enabled researchers to explore different avenues for applications in various fields, from medical to entertainment, pushing the boundaries of what is possible with neurotechnology. The historical trajectory of neurotechnology showcases a remarkable evolution from theoretical concepts to practical applications with real-world impact. From early experiments in controlling cursors to the development of mind-controlled prostheses, BCI technology has demonstrated its potential to revolutionize the way we interact with machines and assist individuals with motor disabilities. Looking forward, the continued advancement of BCIs holds promise for addressing neurological and psychiatric disorders, enhancing communication for individuals with physical limitations, and opening up new possibilities for entertainment and virtual reality experiences. As we stand at the cusp of a new era in human-machine symbiosis, the historical developments of neurotechnology hint at a future where the boundaries between brain and machine will be further blurred, ushering in a new age of innovation and possibility for humanity.

## Early Studies on Brain Function

Early studies on brain function paved the way for the development of BCIs by providing a foundational understanding of how the brain processes information. One significant milestone in neuroscience research was the discovery of neurons and their role in transmitting electrical signals in the brain. This led to the exploration of brain activity and the mapping of different regions responsible for specific functions, such as motor control or sensory perception. Early experiments in the 20th century, like Wilder Penfield's mapping of the human brain using electrical stimulation, laid the groundwork for future advancements in BCI technology by demonstrating the brain's plasticity and adaptability. As neuroscience continued to evolve, researchers began to investigate ways to interact with the brain using external devices, leading to the development of early BCIs. These early studies focused on simple tasks such as controlling cursors on a screen or moving robotic arms using neural signals. By deciphering the brain's electrical activity and translating it into actionable commands, researchers were able to showcase the potential of BCIs in enhancing human-machine interactions. These foundational studies highlighted the feasibility of using neural signals to control external devices, setting the stage for more complex applications in the future. The culmination of early studies on brain function and the development of rudimentary BCIs underscored the potential for revolutionizing human-computer interactions. By deciphering the intricate neural patterns underlying brain function, researchers were able to bridge the gap between the mind and external technology, opening up new possibilities for communication, rehabilitation, and even entertainment. These early endeavors not only laid the groundwork for future advancements in BCI technology but also sparked a wave of innovation that continues to shape the field of neurotechnology today.

## Evolution of BCI Technology

The evolution of BCI technology has been a fascinating journey marked by significant milestones and advancements. From its humble beginnings in the early research and experiments of neuroscience, BCI has now transformed into sophisticated systems capable of translating brain signals into actionable commands. The evolution of BCI technology has been driven by a combination of scientific breakthroughs, technological innovation, and a growing interest in harnessing the power of the human brain to interact with computers and devices in unprecedented ways. As researchers continue to push the boundaries of what is possible with BCI, we are witnessing a shift towards more seamless integration of these interfaces into various aspects of human life. One of the key aspects of BCI technology is its operation, which involves a complex interplay of sensors, signal processing, and decoding algorithms to interpret the neural activity of the brain. Whether through invasive or non-invasive methods, BCIs have the potential to revolutionize the way we interact with technology, opening up new possibilities for individuals with motor disabilities or neurological impairments. The intricate workings of BCI technology are a testament to the interdisciplinary nature of this field, combining elements of neuroscience, engineering, and computer science to create devices that can decode the language of the brain. Looking towards the future, the possibilities offered by BCI technology are both exciting and daunting. As BCIs become more sophisticated and integrated into everyday life, we are faced with ethical and social challenges that must be carefully navigated. Issues surrounding privacy, security, and the potential for technology dependence raise important questions about the impact of BCI on society. With the right approach and consideration, BCI technology has the potential to enhance human capabilities, improve quality of life, and usher in a new era of human-technology interaction that could shape the future of humanity in profound ways.

## Milestones in Neurotechnology Research

One milestone in neurotechnology research was the development of the first BCI system in the 1970s by researchers at the University of California, Los Angeles. This early system allowed individuals to control a computer cursor using brain signals, marking a significant advancement in the field of neural engineering. Subsequent research in the 1980s and 1990s focused on improving the accuracy and speed of BCIs, leading to the development of more sophisticated algorithms for decoding neural signals. These advancements laid the foundation for the commercialization of BCIs and their applications in various industries. Another key milestone in neurotechnology research was the successful demonstration of non-invasive BCIs that could decode brain signals without the need for surgical implants. This breakthrough, achieved in the early 2000s, opened up new possibilities for the widespread adoption of BCIs in healthcare, gaming, and communication. Non-invasive BCIs, such as EEG devices, are now being used in clinical settings to assist patients with severe motor disabilities and in research labs to study brain function. The development of user-friendly and affordable EEG-based BCIs has democratized access to brain-computer interaction and paved the way for innovative applications in the entertainment and consumer tech industries. Recent advancements in neurotechnology research have focused on improving the robustness and reliability of BCIs through the integration of machine learning and AI algorithms. By leveraging these technologies, researchers have been able to develop BCIs that can adapt to individual users' neural patterns, resulting in more precise and responsive interfaces. These technological advances have not only enhanced the performance of BCIs but have also fueled the development of hybrid systems that combine neural inputs with external sensors for greater functionality. The continued evolution of neurotechnology holds great promise for revolutionizing human-computer interaction and enhancing the quality of life for individuals with motor disabilities and neurological conditions.

## III. TYPES OF BCIS

One type of BCI is the non-invasive BCI, which involves using external sensors to detect brain activity without the need for surgical implantation. These sensors, such as EEG devices, can pick up electrical signals produced by the brain and convert them into commands that can be used to control external devices. Non-invasive BCIs are often utilized in research settings, as they are easier to use and less invasive than their invasive counterparts. On the other hand, invasive BCIs require surgical implantation of electrodes directly into the brain tissue, allowing for more precise and detailed measurements of neural activity. These electrodes can pick up individual neuron firing patterns, providing a higher level of control and efficiency compared to non-invasive BCIs. While invasive BCIs offer greater accuracy and potential for advanced applications, they also come with higher risks and ethical considerations due to the invasive nature of the procedure. Hybrid BCIs combine elements of both non-invasive and invasive BCIs, offering a middle ground between accuracy and invasiveness. These systems typically utilize non-invasive sensors to capture brain signals from the surface of the scalp, while also incorporating invasive implants for more precise control and communication with external devices. Hybrid BCIs aim to harness the advantages of both non-invasive and invasive technologies, providing a balance between usability and functionality in various applications, ranging from medical treatments to assistive technologies.

### Invasive BCIs
Invasive BCIs, a subset of BCIs that require surgical implantation

of electrodes directly into the brain tissue, offer unique advantages in terms of signal quality and system performance. By bypassing the scalp and skull, invasive BCIs can access neuronal activity with high spatial and temporal resolution, enabling more precise control of external devices or applications. The ability to record neural signals from individual neurons or small groups of neurons allows for fine-grained decoding of movement intentions or cognitive states, leading to enhanced performance and functionality compared to non-invasive BCIs. Despite their superior performance, invasive BCIs come with inherent risks and challenges that need to be carefully considered. The surgical implantation of electrodes carries the risk of infection, tissue damage, or cognitive side effects, underscoring the importance of proper medical supervision and post-operative care. The long-term stability and reliability of invasive BCIs remain a concern, as neural signals can degrade over time due to tissue response or electrode drift. These factors highlight the need for ongoing research and development to address the technical and clinical challenges associated with invasive BCIs and ensure their safe and effective use in various applications. While invasive BCIs offer unparalleled precision and performance in neural recording and control, their adoption and widespread use are contingent upon addressing the associated risks and challenges. Continued research into materials, surgical techniques, and signal processing algorithms is crucial to enhance the safety, efficacy, and long-term viability of invasive BCIs. By overcoming these obstacles, invasive BCIs have the potential to revolutionize neurotechnology and unlock new possibilities for enhancing human-machine interactions and improving the quality of life for individuals with neurological disorders or disabilities.

## Non-invasive BCIs

The development of non-invasive BCIs has opened up new possibilities for enhancing human-computer interaction without the need for invasive surgical procedures. By utilizing technologies such as EEG and functional near-infrared spectroscopy (fNIRS), researchers have been able to capture neural activity and translate it into commands that can control external devices. This non-invasive approach offers a safer and more accessible option for individuals who may not be suitable candidates for invasive brain implants, thereby democratizing the benefits of BCIs to a wider population. Non-invasive BCIs have shown promise in various applications, ranging from medical interventions to entertainment and communication. In the medical field, non-invasive BCIs have been instrumental in enabling individuals with motor disabilities to regain control over their movements through mind-controlled prostheses or exoskeletons. Non-invasive BCIs have been used in neurorehabilitation programs for stroke patients, helping to improve motor functions and cognitive abilities. These advancements highlight the potential of non-invasive BCIs to revolutionize the field of neurorehabilitation and provide new avenues for enhancing the quality of life for individuals with neurological conditions. Despite the promising applications of non-invasive BCIs, there are still challenges that need to be addressed to fully realize their potential. Issues such as signal quality, accuracy, and reliability continue to be obstacles in the development of robust non-invasive BCIs. Ensuring data privacy and security in the collection and transmission of neural information is paramount to prevent potential breaches or unauthorized access. By addressing these challenges and further refining non-invasive BCI technologies, we can pave the way for a future where seamless communication between the human brain and external devices becomes a reality, transforming the way we interact with technology in our daily lives.

## Hybrid BCIs

One of the most cutting-edge developments in the field of neurotechnology is the emergence of hybrid BCIs, which combine the strengths of different types of BCIs to overcome their individual limitations. By integrating invasive and non-invasive BCI technologies, hybrid BCIs aim to improve signal quality, increase communication speed, and enhance overall performance. This innovative approach holds great promise for a wide range of applications, from medical treatments to entertainment and communication platforms. Through the combination of invasive and non-invasive BCI components, hybrid systems can provide more accurate and reliable neural signals for decoding algorithms. This robust signal processing capability enables more precise control of external devices, such as prosthetic limbs or computer interfaces, leading to improved quality of life for individuals with motor disabilities. The integration of multiple sensor modalities in hybrid BCIs allows for richer information extraction from the brain, opening up new possibilities for advanced neurorehabilitation techniques and therapeutic interventions. The development of hybrid BCIs represents a significant step forward in the evolution of neurotechnology, offering a versatile and adaptable platform for various applications. By leveraging the strengths of both invasive and non-invasive technologies, hybrid BCIs have the potential to revolutionize the way humans interact with machines and the world around them. As research in this area continues to progress, the integration of hybrid BCIs into everyday life could lead to transformative advancements in healthcare, entertainment, and communication, shaping the future of humanity in profound ways.

# IV. APPLICATIONS OF NEUROTECHNOLOGY IN HEALTHCARE

Advancements in neurotechnology have paved the way for innovative applications in healthcare, with BCIs at the forefront of this revolution. BCIs allow direct communication between the brain and external devices, enabling a range of medical interventions that were once only imagined in science fiction. One significant application of neurotechnology in healthcare is the development of mind-controlled prostheses for individuals with motor disabilities. These prostheses can restore lost functionality and independence to users, improving their quality of life and overall well-being. By utilizing BCIs, patients can regain control over their movements through the power of their thoughts, marking a significant milestone in the field of neurorehabilitation. Neurotechnology has shown promise in the treatment of neurological and psychiatric disorders, offering new avenues for therapeutic interventions. BCIs have the potential to assist in treating conditions such as Parkinson's disease, epilepsy, and depression by modulating neural activity and restoring proper brain function. Through real-time monitoring and feedback, BCIs can help clinicians tailor treatments to individual patients, leading to more precise and effective interventions. This personalized approach to healthcare represents a major paradigm shift, moving towards targeted therapies that address the root causes of neurological disorders rather than just managing symptoms. The integration of BCIs in healthcare has opened up possibilities for enhanced communication and interaction for individuals with physical limitations. By utilizing neurotechnology, individuals with severe disabilities can communicate more effectively, express their thoughts, and engage with the world in ways that were previously impossible. BCIs have been instrumental in developing mind-controlled video games and virtual reality experiences, offering entertainment and recreation opportunities for individuals with limited mobility or dexterity. This intersection of technology and healthcare underscores the transformative potential of neurotechnology in improving the quality of life for diverse populations and expanding the boundaries of human capabilities.

## Neurorehabilitation

Neurorehabilitation plays a crucial role in the field of BCIs, as it aims to restore or enhance neural function through technological interventions. This innovative approach utilizes BCI technology to assist individuals with neurological impairments in regaining lost abilities or improving their quality of life. By leveraging the principles of neuroplasticity, neurorehabilitation programs can adapt and optimize BCI systems to suit the specific needs of each patient, facilitating personalized and targeted interventions. Through a combination of neurofeedback, motor imagery, and cognitive training, individuals undergoing neurorehabilitation can actively engage with BCI systems to retrain neural pathways and promote recovery. The integration of neurorehabilitation and BCI technology holds immense promise for enhancing the rehabilitation process for individuals with various neurological conditions, such as stroke or traumatic brain injury. By enabling direct communication between the brain and external devices, BCI systems can augment traditional rehabilitation therapies, offering new avenues for improving motor function, cognitive abilities, and overall well-being. Through real-time feedback and adaptive algorithms, BCI systems can assist patients in performing rehabilitative tasks more effectively, boosting motivation and engagement in the recovery process. This symbiotic relationship between neurorehabilitation and BCI technology opens up exciting possibilities for overcoming physical and cognitive challenges in a more efficient and holistic manner. The synergy between neurorehabilitation and BCI technology signifies a paradigm shift in the approach to neurological rehabilitation, offering new hope and opportunities for individuals with diverse neurological conditions. The personalized and adaptable nature of BCI systems, coupled with the principles of neuroplasticity and targeted interventions, underscores the transformative potential of this integrated approach in improving outcomes and enhancing quality of life for patients. As technological advancements continue to refine and expand the capabilities of BCI systems, the future of neurorehabilitation holds great promise in revolutionizing the field of rehabilitation and empowering individuals to overcome neurological challenges with unprecedented efficiency and efficacy.

## Neuroprosthetics

Neuroprosthetics have the potential to revolutionize the field of medical technology by providing individuals with severe motor disabilities the ability to regain control over their movements. Through the use of BCIs, neuroprosthetics enable patients to operate artificial limbs, wheelchairs, or other assistive devices through direct neural signals. By implanting sensors in the brain that detect neural activity related to movement intentions, BCI technology can decode these signals and translate them into actionable commands for prosthetic devices. This seamless integration of neural control with external technology opens up new possibilities for enhancing the quality of life for individuals with physical limitations. One of the key benefits of neuroprosthetics is the significant improvement in the autonomy and independence of patients with motor disabilities. By bypassing the damaged or impaired neural pathways, BCI technology allows individuals to perform everyday tasks and interact with their environment in ways that were previously impossible. This newfound freedom can have a profound impact on the mental and emotional well-being of patients, empowering them to lead more fulfilling and active lives. Neuroprosthetics have the potential to enable individuals to regain a sense of agency and control over their bodies, restoring a level of functionality that was once thought to be lost. The development of neuroprosthetics represents a remarkable intersection of neuroscience, engineering, and medical innovation, with the potential to transform the way we perceive and interact with technology. As the field of neurotechnology continues to advance, the future of neuroprosthetics holds promise for further enhancements in functionality, dexterity, and adaptability. By addressing the current challenges and limitations of BCI technology, researchers can pave the way for more seamless and intuitive integration of neuroprosthetics into everyday life. As we navigate the ethical and societal implications of this groundbreaking technology, it is essential to remain vigilant in ensuring that the benefits of neuroprosthetics are equitably distributed and that the autonomy and privacy of individuals are safeguarded. The future of neuroprosthetics holds immense potential for transforming the lives of individuals with motor disabilities and reshaping the landscape of assistive technology.

## Neural Imaging and Diagnosis

Advancements in neural imaging have revolutionized the field of diagnosis, allowing for more accurate and timely identification of various neurological conditions. Techniques such as functional magnetic resonance imaging (fMRI) and positron emission tomography (PET) provide valuable insights into brain activity and can help differentiate between different disorders. By analyzing blood flow and metabolic activity in specific brain regions, clinicians can diagnose conditions such as Alzheimer's disease, stroke, or epilepsy with greater precision. Neural imaging has also enabled the early detection of conditions like brain tumors, leading to faster intervention and improved outcomes for patients. Neural imaging techniques have enhanced the understanding of the underlying mechanisms of certain neurological disorders, facilitating more targeted treatment strategies. Diffusion tensor imaging (DTI) allows for the visualization of white matter tracts in the brain, aiding in the diagnosis of conditions like multiple sclerosis or traumatic brain injury. By identifying structural abnormalities or disruptions in neural pathways, clinicians can tailor treatment plans to address specific issues, improving the overall efficacy of interventions. Neural imaging has therefore not only improved diagnostic accuracy but also paved the way for personalized and precise medical interventions in neurology. In addition to clinical applications, neural imaging plays a crucial role in research and the development of new diagnostic tools and treatments for neurological disorders. By providing detailed insights into brain function and structure, imaging techniques contribute to the identification of biomarkers and potential therapeutic targets. Researchers can use neural imaging to monitor the progression of diseases, evaluate the effectiveness of interventions, and assess the impact of novel treatments. The integration of advanced imaging technologies with AI and machine learning algorithms further enhances the diagnostic capabilities in neurology, enabling more accurate and efficient identification of neurological conditions. Neural imaging continues to drive innovation in diagnosis and treatment, offering new horizons for personalized medicine and improved patient care.

# V. ETHICAL CONSIDERATIONS IN NEUROTECHNOLOGY

One of the critical aspects to consider when discussing neurotechnology, particularly BCIs, is the ethical implications that arise from direct manipulation of the human brain. As technology advances and the lines between man and machine blur, questions surrounding privacy, autonomy, and consent become increasingly pertinent. The use of BCIs raises concerns about the potential exploitation of neural data, as well as the risks of unauthorized access to an individual's innermost thoughts and feelings. The idea of enhancing cognitive abilities or altering emotional states through direct neural intervention opens up a Pandora's box of ethical dilemmas, challenging our very notions of identity and agency. In addition to privacy concerns, the reliance on neurotechnology for everyday activities poses a significant ethical challenge in terms of technology dependence and potential abuse. As individuals become increasingly reliant on BCIs for communication, entertainment, and even medical purposes, the risk of losing control over one's own thoughts and actions looms large. The unequal access to advanced neurotechnologies could exacerbate existing social inequalities, creating a divide between those who can afford cutting-edge enhancements and those who are left behind. Addressing these ethical dilemmas is crucial to ensuring that the benefits of neurotechnology are equitably distributed and do not inadvertently harm vulnerable populations. Despite the ethical challenges that neurotechnology presents, there is also immense potential for enhancing the quality of human life and opening up new possibilities for humanity. By engaging in thoughtful dialogue and establishing robust ethical frameworks, we can harness the transformative power of BCIs to improve healthcare, empower individuals with disabilities, and even expand our understanding of consciousness and cognition. By navigating the complex terrain of ethical considerations in neurotechnology, we can shape a future where technology serves as a tool for human flourishing rather than a source of division and harm.

## Privacy and Data Security

Privacy and data security are paramount concerns in the realm of BCIs. As these technologies advance and become more integrated into everyday life, the potential for sensitive neural data to be intercepted or misused raises significant ethical and legal questions. The very nature of BCIs involves direct access to an individual's brain activity, raising concerns about the protection of personal thoughts and information. Ensuring robust encryption and secure storage of neural data is essential to safeguard the privacy of users and prevent unauthorized access. The interconnectedness of BCIs with external devices and networks introduces vulnerabilities that can be exploited by malicious actors. As BCIs continue to evolve and become more interconnected with the internet of things (IoT), the risk of cyberattacks targeting neural interfaces grows. Ensuring the integrity and security of data transmission between the brain and external devices is crucial to prevent unauthorized access and potential manipulation of neural signals. Implementing stringent data encryption protocols and security measures is essential to protect users from privacy breaches and data theft. The ethical implications of data security in BCI technology extend beyond individual privacy to broader societal concerns. The potential for mass surveillance and the exploitation of neural data for commercial or political purposes raise ethical dilemmas regarding consent, autonomy, and the rights of individuals. As BCI technology becomes more pervasive in various sectors, including healthcare, gaming, and communication, establishing clear guidelines and regulations to protect privacy and data security is imperative. Striking a balance between innovation and ethical considerations will be essential to harness the full potential of BCIs while safeguarding privacy and upholding ethical standards.

## Informed Consent and Autonomy

One of the key ethical considerations in the utilization of BCIs is the issue of informed consent and autonomy. Informed consent refers to the process in which individuals are provided with relevant information about a particular procedure or technology and are given the opportunity to make a voluntary decision about whether to participate or not. This is particularly crucial in the context of BCI, as the technology involves direct access to an individual's brain activity and has the potential to influence their thoughts and actions. Ensuring that individuals fully understand the implications of using BCI and have the autonomy to make informed choices is essential in upholding their rights and dignity. The concept of autonomy plays a significant role in the ethical considerations surrounding BCI. Autonomy refers to an individual's ability to make decisions and take actions based on their own values and beliefs, free from undue influence or coercion. When it comes to BCI, the technology has the capacity to impact not only a person's physical capabilities but also their cognitive and emotional processes. As such, safeguarding autonomy becomes paramount, as individuals must have the freedom to control how their neural data is collected, used, and shared. This ensures that individuals retain agency over their own thoughts and behaviors, even in the presence of advanced technology that can interface directly with the brain. Navigating the complex terrain of informed consent and autonomy in the realm of BCI requires a delicate balance between promoting innovation and safeguarding individual rights. By establishing clear guidelines for obtaining informed consent and protecting autonomy, we can create a framework that respects the dignity and self-determination of individuals using BCI technology. As we continue to explore the potential of BCIs in enhancing human capabilities, it is imperative that we prioritize ethical considerations to ensure that the benefits of the technology are maximized while minimizing potential risks and infringements on individual autonomy.

## Equity and Access to Neurotechnological Advancements

As neurotechnological advancements continue to push boundaries, the issue of equity and access becomes increasingly pertinent. The development of BCIs has the potential to revolutionize human-machine interactions, offering new opportunities for individuals with physical limitations or neurological impairments. The distribution of these advancements is not equal, raising concerns about accessibility and fairness. As with many emerging technologies, there is a risk that only certain privileged groups will benefit from BCIs, while others are left behind. This highlights the importance of addressing equity in the development and deployment of neurotechnological advancements to ensure that the benefits are shared equitably among all members of society. Ensuring equitable access to BCIs involves considering not only physical access but also issues of affordability, usability, and inclusivity. People from marginalized communities may face barriers such as cost, lack of technical expertise, or cultural differences that prevent them from fully benefiting from neurotechnological advancements. It is crucial to address these disparities by designing BCIs with diverse user needs in mind, conducting outreach programs to educate underserved populations, and implementing policies that promote inclusivity and accessibility. By actively engaging with a wide range of stakeholders, including those from underrepresented communities, developers can ensure that BCIs are designed and distributed in a way that promotes equity and social justice. The ethical imperative of ensuring equity and access to neurotechnological advancements cannot be overstated. As BCIs become more integrated into our daily lives, it is essential to consider how these technologies may impact different populations and to strive for inclusivity and fairness in their development and deployment. By addressing the challenges of equity and access head-on, we can work towards a future where neurotechnological advancements benefit all members of society, regardless of their background or circumstances. Only through a concerted effort to promote equity and access can we harness the full potential of BCIs to improve the quality of life for humanity as a whole.

# VI. NEUROTECHNOLOGY IN COGNITIVE ENHANCEMENT

Neurotechnology has paved the way for cognitive enhancement through the development of BCIs, offering promising avenues for revolutionizing human interaction with technology. As these technologies continue to evolve, their potential to augment cognitive abilities holds significant implications for various fields, from healthcare to entertainment. The seamless integration of BCIs into everyday life could offer unprecedented opportunities for enhancing communication, mobility, and overall quality of life for individuals with physical limitations or neurological disorders. One of the most compelling aspects of neurotechnology in cognitive enhancement is its application in the medical and therapeutic realm. BCIs have shown great potential in the treatment of motor disabilities, enabling individuals to control prosthetic limbs and devices through direct neural signals. These interfaces have demonstrated effectiveness in neurorehabilitation for stroke patients, offering new hope for restoring lost motor functions. The ability of BCIs to directly interface with the brain opens up promising avenues for treating a wide range of neurological and psychiatric disorders, potentially revolutionizing the field of neuromedicine. Beyond the realm of healthcare, the integration of BCIs in entertainment and communication presents exciting possibilities for enhancing human experience. From mind-controlled video games to immersive virtual reality experiences, these technologies offer new ways to engage with entertainment media and create interactive experiences. BCIs have the potential to revolutionize communication for individuals with physical limitations, offering enhanced means of expressing thoughts and ideas. As neurotechnology continues to advance, the boundaries between human cognition and technological interfaces may blur, opening up new horizons for cognitive enhancement and human-machine interaction.

## Memory Augmentation

As memory augmentation technologies continue to advance, concerns have been raised regarding the potential ethical implications of accessing, storing, and manipulating memories. The ability to alter or delete memories raises questions about consent, autonomy, and the potential for abuse. In a society where memories are increasingly becoming digitized, questions about ownership and privacy rights also come to the forefront. It is crucial to establish robust ethical guidelines and regulations to govern the use of memory augmentation technologies to ensure that individuals' rights and well-being are protected. The integration of memory augmentation technologies into everyday life raises important societal considerations. The ability to enhance cognitive abilities through memory augmentation may create a divide between those who have access to such technologies and those who do not. This could exacerbate existing inequalities and lead to a new form of cognitive elitism. The impact of relying on external devices for memory storage and retrieval may affect individuals' ability to form meaningful connections and engage in critical thinking. It is essential to carefully consider the social implications of widespread adoption of memory augmentation technologies to foster a more equitable and inclusive society. Memory augmentation holds the promise of revolutionizing how we interact with and utilize our cognitive capabilities. From enhancing learning and cognitive performance to providing personalized assistance for individuals with memory impairments, the potential applications of memory augmentation technologies are vast. As with any technological advancement, it is crucial to approach the development and implementation of these technologies with a critical eye towards ethical considerations, societal impacts, and the overall well-being of individuals. By navigating these challenges thoughtfully and responsibly, memory augmentation has the potential to significantly improve the quality of life for individuals and open up new possibilities for humanity as a whole.

## Cognitive Skills Training

As cognitive skills training continues to gain traction in the field of neuroscience, researchers are exploring innovative methods to enhance various aspects of cognitive function. One approach that has garnered significant attention is the use of BCIs to target specific cognitive abilities. By leveraging the principles of neuroplasticity, BCIs can be designed to facilitate cognitive training through real-time feedback and interactive tasks. These training programs can be tailored to address different cognitive domains such as memory, attention, and executive functions, providing a personalized and adaptive learning experience for individuals seeking to improve their cognitive abilities. The integration of cognitive skills training with BCIs holds promise for enhancing cognitive performance in diverse populations, including individuals with neurological disorders, aging adults, and even healthy individuals looking to optimize their mental capabilities. By harnessing the power of neurotechnology, individuals can engage in targeted cognitive training exercises that stimulate neural networks associated with specific cognitive functions. This targeted approach not only enhances cognitive skills but also promotes neuroplasticity and neural rehabilitation, leading to long-lasting improvements in cognitive performance. The combination of cognitive skills training with BCIs opens up new avenues for exploring the potential of neural enhancement and cognitive augmentation. As technology continues to advance, researchers are pushing the boundaries of what is possible in terms of optimizing cognitive function through neurofeedback and brain-computer interactions. By harnessing the adaptive capabilities of the brain and leveraging cutting-edge technology, individuals may soon be able to unlock new levels of cognitive potential, revolutionizing the way we think about cognition and human intelligence. With further research and development in this field, the future of cognitive skills training with BCIs holds immense promise for enhancing cognitive abilities and unlocking the full potential of the human brain.

## Ethical Implications of Cognitive Enhancement

As the field of cognitive enhancement continues to advance, ethical considerations surrounding the use of technologies like BCIs become increasingly important. One key ethical implication is the potential for inequality and social division to arise from unequal access to cognitive enhancement technologies. If only a privileged few have access to technologies that enhance cognitive abilities, it could exacerbate existing societal disparities and create a new form of discrimination based on cognitive enhancement status. This raises questions about fairness, justice, and the distribution of resources in a future where cognitive enhancement is widespread. Another ethical concern related to cognitive enhancement through BCIs is the issue of autonomy and personal identity. By directly manipulating brain function to enhance cognitive abilities, individuals may risk losing a sense of autonomy and authenticity. The use of BCIs to improve cognitive function raises questions about the true nature of selfhood and whether individuals can truly claim credit for their enhanced cognitive abilities if they are artificially augmented. This blurring of the lines between natural and enhanced cognitive abilities raises important questions about personal identity and the nature of humanity in a world where cognitive enhancement technologies are ubiquitous. The ethical implications of cognitive enhancement technologies extend to issues of safety, privacy, and consent. As BCIs become more sophisticated and capable of directly interfacing with the human brain, concerns about the potential for abuse, hacking, and unauthorized access to neural data become more pressing. Safeguarding the privacy and security of neural data collected and processed by BCIs is crucial to ensuring the ethical use of these technologies. Obtaining informed consent from individuals who choose to undergo cognitive enhancement procedures with BCIs is essential to upholding principles of autonomy and respect for individual rights. Addressing these ethical challenges will be essential to realizing the full potential of cognitive enhancement technologies while minimizing potential harms and risks to society.

# VII. NEUROTECHNOLOGY AND HUMAN-MACHINE INTERACTION

Neurotechnology has made significant advancements in recent years, leading to the development of BCIs that have the potential to revolutionize human-machine interaction. These interfaces bridge the gap between the human brain and external devices, allowing for seamless communication and control. The evolution of this technology from its early stages to the sophisticated systems of today is a testament to the progress made in neuroscience and engineering. Researchers have delved deep into understanding the complexities of the brain, paving the way for new possibilities in areas such as medical treatment, communication, and entertainment. One of the key components of BCIs is the intricate process through which neural signals are translated into actionable commands for external devices. By utilizing sensors, signal processing techniques, and decoding algorithms, BCIs enable individuals to control technology through their thoughts alone. Different types of BCIs, such as invasive and non-invasive interfaces, offer varying levels of precision and accessibility to users. With the continued advancements in this technology, the potential applications in various fields, including healthcare, gaming, and communication, are continually expanding. The integration of BCIs into everyday life presents both exciting opportunities and ethical challenges. While BCIs hold the promise of improving quality of life for individuals with disabilities and enhancing human potential, concerns about privacy, security, and the ethical implications of manipulating neural data persist. As we navigate this brave new world of neurotechnology, it is imperative to address these ethical and societal issues to ensure that the benefits of BCIs can be maximized while minimizing potential risks. The future of humanity may be shaped by the co-evolution of humans and technology, with BCIs playing a pivotal role in this transformative journey.

## Augmented Reality and Virtual Reality

As augmented reality and virtual reality technologies continue to advance, they are poised to revolutionize the way humans interact with the digital world. AR enhances the real world by overlaying digital information on it, while VR creates immersive, artificial environments. Both technologies have a wide range of applications, from entertainment and gaming to education, healthcare, and industrial training. The adoption of AR and VR is steadily increasing, with companies investing in developing AR and VR platforms for various industries. One of the key advantages of AR and VR is their ability to provide immersive experiences that can enhance learning and training. In education, AR and VR can transport students to historical events, outer space, or microscopic worlds, enhancing engagement and knowledge retention. In healthcare, surgeons can use AR to visualize medical imaging data in real time during procedures, improving accuracy and patient outcomes. AR and VR can help individuals with disabilities by providing simulated environments for therapy and rehabilitation, unlocking new possibilities for personalized care. Despite their many benefits, AR and VR technologies also present challenges that must be addressed. Issues such as privacy concerns, ethical considerations, and potential negative effects on mental health need to be carefully managed as these technologies become more widespread. The cost and accessibility of AR and VR hardware remain barriers to mass adoption. As these technologies continue to evolve, stakeholders must work together to ensure that AR and VR are developed in a responsible and sustainable manner, maximizing their potential to improve human experiences in the digital age.

## Neurofeedback in Gaming

Neurofeedback in gaming has recently gained considerable attention as a promising approach to enhancing the gaming experience. By utilizing BCIs, players can interact with games using their brain activity, opening up a new realm of possibilities for game developers and players alike. Neurofeedback systems allow for real-time monitoring of brain activity, providing valuable insights into the cognitive and emotional states of the player. This data can be used to adapt the gameplay experience, creating more immersive and personalized gaming experiences. One of the key advantages of incorporating neurofeedback in gaming is its potential to improve cognitive abilities and emotional regulation in players. By providing feedback on brain activity, players can learn to better control their cognitive processes, such as attention and memory, leading to enhanced performance in games. Neurofeedback can help in managing stress and anxiety levels during gameplay, promoting a more enjoyable and fulfilling gaming experience. This aspect of neurofeedback in gaming has significant implications for the future development of games that can adapt to the player's mental state in real-time. The integration of neurofeedback in gaming holds promise for enhancing the overall well-being of players by promoting mindfulness and self-awareness. Through the use of BCIs, players can engage in games that encourage relaxation, focus, and emotional regulation, fostering a sense of mental well-being. This novel approach to gaming not only offers entertainment value but also has the potential to contribute to the broader field of mental health and wellness. As neurofeedback technology continues to evolve, the possibilities for leveraging brain activity in gaming are vast, opening up new horizons for the future of interactive entertainment.

## Implications for Education and Training

As the field of neurotechnology continues to advance, the implications for education and training are becoming increasingly significant. BCIs have the potential to revolutionize the way students learn and interact with educational material. By directly interfacing with the brain, BCIs can provide real-time feedback on a student's understanding and engagement, allowing for personalized learning experiences tailored to individual needs and abilities. This adaptability could lead to more efficient and effective learning outcomes, as educators can adjust their teaching methods based on the data gathered from BCIs. The integration of BCIs in educational settings could enhance accessibility for individuals with disabilities, allowing them to fully participate in learning activities that were previously inaccessible. Students with physical disabilities that affect their ability to communicate or manipulate traditional learning tools could use BCIs to interact with educational software and resources in a more efficient and independent manner. This inclusivity can create a more equitable educational environment, ensuring that all students have the opportunity to thrive and succeed academically. In addition to improving the learning experience, BCIs can also be utilized in training programs and professional development. In fields such as healthcare or aviation, where quick decision-making and precise actions are crucial, BCIs can be used to provide real-time feedback on an individual's cognitive and physical performance. This level of data-driven training can help individuals hone their skills and improve their overall performance in high-stakes situations. The integration of BCIs in education and training holds immense potential to enhance learning outcomes, promote inclusivity, and optimize professional development in various fields.

# VIII. NEUROTECHNOLOGY AND NEUROETHICS

As neurotechnology continues to advance, the field of neuroethics is also gaining prominent attention. Ethical considerations surrounding the use of BCIs have become a crucial aspect of discussions on the future of neurotechnology. Issues such as privacy and security of neural data, the potential risks of technology dependence and abuse, and the ethical implications of directly manipulating the human brain are complex and multifaceted. These ethical dilemmas must be carefully addressed to ensure that the benefits of BCI technology are maximized while minimizing potential harms to individuals and society as a whole. The integration of BCIs into everyday life raises important questions about the impact on society and the co-evolution of humans and technology. As these technologies become more pervasive, it is essential to consider how they may influence social dynamics, interpersonal relationships, and even the concept of personal identity. The transformative potential of BCIs to improve the quality of life for individuals with disabilities and neurological disorders is undeniable. Careful consideration must be given to the broader societal implications of widespread BCI adoption, including issues related to equity, access, and the potential for exacerbating existing inequalities. While the future of BCIs holds great promise for enhancing human capabilities and quality of life, it is crucial to approach their development and implementation with careful consideration of the ethical and social challenges they present. By fostering interdisciplinary collaborations between neuroscientists, engineers, ethicists, policymakers, and other stakeholders, we can ensure that neurotechnology is developed and used in a manner that aligns with ethical principles, respects individual rights, and promotes social good. By navigating these challenges thoughtfully, we can strive towards a future where neurotechnology enhances human potential in ways that are ethical, equitable, and sustainable.

## Neurodiversity and Inclusivity

The concept of neurodiversity emphasizes the idea that neurological differences, such as those found in people with autism, ADHD, or dyslexia, should be recognized and respected as part of the natural variation in human cognition. Inclusivity within the context of neurodiversity involves creating environments that accommodate and celebrate these differences, rather than trying to force individuals to conform to neurotypical standards. By embracing neurodiversity, organizations can foster creativity, innovation, and problem-solving skills that may not be present in a homogenous group. One key aspect of promoting inclusivity within neurodiversity is providing reasonable accommodations and support for individuals with diverse neurological profiles. This can include flexible work arrangements, sensory-friendly spaces, clear communication strategies, and training programs for neurotypical colleagues to better understand and support their neurodivergent peers. By making these accommodations, organizations can create a more inclusive and supportive work environment where all individuals can thrive and contribute their unique perspectives and talents. In order to truly promote inclusivity within neurodiversity, it is essential for organizations to adopt a culture of acceptance, understanding, and empathy towards individuals with diverse neurological profiles. This involves challenging stereotypes and biases, promoting education and awareness about neurodiversity, and actively seeking out and valuing the contributions of neurodivergent individuals. By fostering a culture of inclusivity and understanding, organizations can create a more equitable and diverse workforce that harnesses the full potential of all individuals, regardless of their neurological differences. Embracing neurodiversity and inclusivity can lead to a more vibrant, innovative, and successful organization that benefits from the unique talents and perspectives of all its members.

## Neuroexistentialism and Identity

Neuroexistentialism and identity play a significant role in the discourse surrounding BCIs and neurotechnology. Neuroexistentialism, a philosophical concept that examines the intersection of neuroscience and existential questions about human identity and consciousness, becomes particularly relevant in the context of BCI. As individuals start to navigate the integration of technology with their own cognitive processes, questions about the impact on personal identity and the nature of human existence inevitably arise. The ability to control external devices or communicate through thoughts challenges traditional notions of self and raises ethical dilemmas about the authenticity of human agency in a technologically mediated world. The concept of identity is further complicated by the potential for BCI to alter or enhance cognitive abilities, prompting discussions about the boundaries of human enhancement and the implications for individual autonomy. Neuroexistentialist thinkers argue that the integration of technology into our brains blurs the distinction between the self and external tools, raising concerns about the authenticity and integrity of personal identity in a digitally augmented reality. The ability of BCIs to access and manipulate neural activity raises questions about privacy, consent, and the ethical implications of directly influencing brain function. As researchers and developers continue to advance BCI technology, it is crucial to consider the existential and identity-related implications of these innovations. By addressing the neuroexistential aspects of BCI, we can better understand the complex interplay between technology and human identity, paving the way for ethical and responsible implementation of neurotechnology. Exploring the philosophical dimensions of BCI through the lens of neuroexistentialism can enrich our understanding of the human experience and guide us toward a more thoughtful and nuanced approach to integrating technology into our lives.

## Regulation and Governance of Neurotechnologies

As neurotechnologies continue to advance, the need for effective regulation and governance becomes paramount. The intricacies of BCIs require careful oversight to ensure ethical use and mitigate potential risks. Governments, regulatory bodies, and industry leaders must collaborate to establish frameworks that address data privacy, security, and accountability in the development and deployment of neurotechnologies. By implementing robust regulations, stakeholders can foster innovation while safeguarding the well-being of individuals who interact with BCIs. One approach to regulating neurotechnologies involves setting standards for data collection, storage, and sharing. Given the sensitive nature of neural data, it is essential to establish guidelines that protect user privacy and prevent unauthorized access. Transparency in the design and functionality of BCIs is crucial for building trust among users and stakeholders. Clear regulations can help mitigate concerns surrounding the misuse of neural information and ensure that neurotechnologies are developed and utilized ethically. Governance mechanisms should address the equitable distribution of benefits and risks associated with neurotechnologies. As BCIs hold the potential to revolutionize healthcare, communication, and entertainment, it is vital to consider the societal implications of their widespread adoption. Stakeholders must engage in dialogue to assess the social impact of BCIs and implement policies that promote inclusivity and accessibility. By prioritizing the well-being of individuals and communities, regulatory frameworks can guide the responsible development and deployment of neurotechnologies for the betterment of society.

# IX. CHALLENGES AND FUTURE DIRECTIONS IN NEUROTECHNOLOGY

One of the key challenges facing the field of neurotechnology is the development of more advanced and accurate BCIs. While significant progress has been made in recent years, there are still limitations in terms of the speed, accuracy, and reliability of current BCIs. Future research directions in this area will focus on improving signal processing techniques, enhancing the resolution of neural signals, and exploring new methods for decoding brain activity. These advancements will be crucial for expanding the potential applications of BCIs beyond the medical field and into areas such as communication, entertainment, and cognitive enhancement. Another major challenge in neurotechnology is ensuring the ethical and societal implications are carefully considered as the technology continues to advance. Issues of privacy, security, and consent must be addressed to safeguard the rights and autonomy of individuals using BCIs. Concerns about technology dependence, potential abuse, and the ethical implications of directly manipulating the human brain will need to be carefully navigated. Future research and development in neurotechnology must be accompanied by robust ethical frameworks and regulatory guidelines to ensure that the technology is used responsibly and ethically. Looking towards the future, the field of neurotechnology holds immense promise for revolutionizing human-machine interaction and enhancing the quality of life for individuals with disabilities. As the capabilities of BCIs continue to improve and expand, we can expect to see more widespread adoption of the technology in various domains. From mind-controlled prosthetics to immersive virtual reality experiences, the potential applications of BCIs are vast and varied. By addressing current challenges and embracing future opportunities, neurotechnology has the potential to transform not only how we interact with technology, but also how we understand and harness the power of the human brain.

## Technological Limitations and Innovations

As neurotechnology continues to advance, researchers are faced with both technological limitations and opportunities for innovation. One of the primary challenges lies in the development of more sophisticated sensors that can accurately capture and interpret neural signals. This is crucial for enhancing the speed and accuracy of BCIs in various applications, such as controlling prosthetic limbs or accessing information directly from the brain. Improving the decoding algorithms used in BCIs can help in translating neural activity into meaningful commands, ultimately enhancing the user experience and expanding the potential applications of these devices. Innovations in signal processing are also key to overcoming technological limitations in neurotechnology. As BCIs generate vast amounts of data from neural signals, efficient processing methods are essential for extracting relevant information and translating it into actionable commands or feedback. Advances in machine learning and AI have shown promise in improving the performance of BCIs by enabling real-time data analysis and adaptive signal processing. By harnessing these innovative approaches, researchers can enhance the capabilities of BCIs and open up new possibilities in fields such as healthcare, communication, and entertainment. The integration of novel materials and technologies in BCIs presents an exciting avenue for technological innovation.

The use of flexible and biocompatible materials in neural interfaces can enhance the longevity and reliability of implantable devices, reducing the risk of tissue damage or immune rejection. The incorporation of wireless communication technologies can enable more seamless interactions between the brain and external devices, improving the overall user experience and mobility. By overcoming these technological limitations through continuous innovation, BCIs hold immense potential to revolutionize how humans interact with technology and enhance the quality of life for individuals with physical disabilities or neurological disorders.

## Ethical and Social Implications

In considering the ethical and social implications of BCIs, one of the primary concerns revolves around the privacy and security of neural data. As BCIs directly interface with the brain, the potential for accessing sensitive information raises significant privacy issues. Unauthorized access to neural data could lead to violations of personal autonomy and the potential for manipulation. The security of neural data stored or transmitted through BCIs must be robust to prevent data breaches and unauthorized use. Ensuring the protection of individuals' neural information is crucial in the development and implementation of BCI technology to safeguard against potential exploitation. Another ethical challenge posed by BCIs is the risk of technology dependence and abuse. As individuals become increasingly reliant on BCIs for communication, control of devices, or even decision-making processes, there is a concern that over-reliance on this technology may lead to a loss of autonomy and agency. The potential for misuse or abuse of BCIs, such as manipulating individuals' thoughts or actions, highlights the importance of establishing clear ethical guidelines and regulations. Addressing these concerns requires a careful balance between harnessing the benefits of BCI technology and mitigating the risks associated with its misuse. The ethical implications of direct manipulation of the human brain through BCIs raise complex moral questions about the boundaries of human identity and agency. The ability to directly interface with the brain to enhance cognitive abilities or alter emotional states raises concerns about the potential for unintended consequences and ethical dilemmas. Questions regarding autonomy, consent, and the implications of altering fundamental aspects of human cognition and behavior necessitate thoughtful consideration in the development and deployment of BCI technology. Balancing the potential benefits of enhanced cognitive abilities with the ethical responsibilities to respect individual autonomy and dignity presents a central challenge in navigating the ethical landscape of BCIs.

## Interdisciplinary Collaboration in Advancing Neurotechnology

Interdisciplinary collaboration plays a crucial role in advancing neurotechnology, particularly in the development of BCIs. By bringing together experts from diverse fields such as neuroscience, engineering, computer science, and psychology, researchers can leverage their unique perspectives and knowledge to overcome complex challenges in BCI technology. Neuroscientists can provide insights into the functioning of the brain, engineers can design innovative hardware and software solutions, and psychologists can contribute their expertise in human behavior and cognition. This interdisciplinary approach allows for a comprehensive understanding of the brain and its interactions with technology, leading to more efficient and effective advancements in neurotechnology. Interdisciplinary collaboration fosters creativity and innovation in the field of neurotechnology. By breaking down traditional silos and encouraging collaboration across disciplines, researchers can explore new ideas and approaches that may not have been possible within a single discipline. This multidisciplinary approach enables researchers to think outside the box, pushing the boundaries of what is currently possible in BCI technology. The integration of machine learning algorithms from computer science with neural signal processing techniques from neuroscience has led to significant improvements in the accuracy and speed of BCIs. This cross-pollination of ideas and expertise ultimately results in more robust and cutting-edge neurotechnological developments. Interdisciplinary collaboration is essential for the advancement of neurotechnology and the realization of the full potential of BCIs. By harnessing the collective expertise of researchers from various disciplines, we can address the complex challenges inherent in developing BCIs and pave the way for groundbreaking innovations in the field. Moving forward, fostering collaborations between neuroscientists, engineers, computer scientists, and psychologists will be crucial for pushing the boundaries of neurotechnology, improving the lives of individuals with neurological disorders, and unlocking new possibilities for human-machine interactions.

# X. NEUROTECHNOLOGY AND NEUROAESTHETICS

Neurotechnology and neuroaesthetics are two interdisciplinary fields that have the potential to revolutionize our understanding of art, perception, and cognition. Neuroaesthetics, which explores the neural basis of beauty and aesthetic experiences, can benefit significantly from advancements in neurotechnology such as BCIs. By using BCIs to monitor brain activity in response to visual stimuli, researchers can gain insight into how individuals process aesthetic information and make subjective judgments about art. This neuroscientific approach to aesthetics allows for a deeper understanding of the cognitive processes underlying aesthetic preferences and can also inform the creation of more engaging and impactful artistic experiences. The integration of neurotechnology and neuroaesthetics has practical applications in fields such as art therapy, design, and marketing. BCIs can be used to tailor experiences based on individual aesthetic preferences, leading to more personalized art interventions and therapeutic interventions. In design and marketing, neuroaesthetic insights can help companies create products and campaigns that are visually appealing and emotionally resonant. By leveraging neurotechnology to decode neural responses to different aesthetic stimuli, businesses can optimize their designs and advertisements to better capture consumers' attention and evoke positive emotional responses. The synergy between neurotechnology and neuroaesthetics holds great promise for advancing our understanding of human perception, creativity, and emotional responses to art. As BCIs continue to evolve and become more sophisticated, researchers will have unprecedented tools to unravel the mysteries of aesthetic experience and develop innovative applications that enhance our interactions with art and design. By harnessing the power of neurotechnology in the realm of aesthetics, we can unlock new possibilities for creative expression, emotional engagement, and personal growth.

## Impact of Neurotechnology on Artistic Expression

The impact of neurotechnology on artistic expression goes beyond traditional forms of art creation, opening up new possibilities for artists to engage with their audience on a deeper level. Through the use of BCIs, artists can now create interactive and immersive experiences that blur the lines between viewer and creator. By harnessing the power of neurotechnology, artists can tap into the subconscious mind of their audience, eliciting emotional responses and connections that were previously unattainable. This level of engagement and interactivity can result in a more profound and meaningful experience for both the artist and the viewer. One of the key ways in which neurotechnology is transforming artistic expression is through the creation of neurofeedback art, where the brain's electrical activity is used to generate visual or auditory outputs. This form of art not only allows for the exploration of the inner workings of the mind but also challenges the conventional boundaries of creativity and expression. By directly interfacing with the brain, artists can delve into the neural processes that underpin our thoughts, emotions, and perceptions, resulting in a more authentic and visceral form of artistic output. This fusion of art and neuroscience has the potential to revolutionize the way we understand and engage with artistic works, paving the way for a new era of creative expression. The integration of neurotechnology into the artistic process can lead to a greater sense of accessibility and inclusivity within the art world. By enabling individuals with disabilities to express themselves through art in novel ways, BCIs can break down barriers and empower marginalized communities to participate in the creation and appreciation of art. This democratization of the artistic landscape not only enriches the cultural tapestry of society but also fosters a more inclusive and diverse artistic community. As neurotechnology continues to advance, the potential for innovation and collaboration within the realm of artistic expression is limitless, offering new avenues for exploration, experimentation, and creativity.

## Neuroaesthetic Experiences in Virtual Reality

Virtual reality has significantly expanded the possibilities for neuroaesthetic experiences, offering a unique platform for exploring the intersection of art, technology, and the brain. By immersing users in digital environments, VR can elicit strong emotional responses and stimulate cognitive processes associated with aesthetics. Studies have shown that engaging with art in VR can activate regions of the brain involved in reward, pleasure, and emotional processing, enhancing the overall aesthetic experience. This digital medium allows for the creation of interactive and multisensory artworks that can provoke intense feelings of presence and immersion, leading to a heightened sense of aesthetic engagement. Neuroaesthetic experiences in VR can provide insights into the neural mechanisms underlying aesthetic perception and appreciation. By monitoring brain activity through advanced neuroimaging techniques such as fMRI and EEG while users engage with VR artworks, researchers can analyze how the brain responds to different visual stimuli and aesthetic properties. This neuroscientific approach can help unravel the complex interplay between sensory perception, emotional processing, and cognitive evaluation in shaping aesthetic experiences. Understanding these neural correlates of aesthetic perception in VR can not only enhance our appreciation of art but also inform the design of more compelling and impactful virtual experiences. The integration of neuroaesthetics and VR holds immense potential for expanding the boundaries of artistic expression and creating transformative experiences for users. By leveraging insights from neuroscience and digital technology, artists and designers can push the boundaries of creativity and innovation in the digital realm. The immersive and interactive nature of VR can democratize access to art and culture, allowing individuals from diverse backgrounds to engage with and appreciate aesthetic experiences in new and meaningful ways. As technology continues to evolve, the fusion of neuroaesthetics and VR promises to redefine our understanding of art, perception, and the human experience in unprecedented ways.

## Neuroaesthetics in Design and Architecture

One fascinating application of neurotechnology is its integration into design and architecture, giving birth to the field of neuroaesthetics. By leveraging our understanding of how the brain perceives and responds to aesthetics, designers and architects can create spaces and structures that evoke specific emotional and cognitive responses. Neuroaesthetics in design and architecture not only enhances the visual appeal of buildings but also promotes well-being and productivity in occupants. This interdisciplinary approach allows for the creation of spaces that are not only visually pleasing but also cognitively stimulating, ultimately improving the overall experience for individuals interacting with the environment. Through the lens of neuroaesthetics, designers can harness the power of elements such as symmetry, color, lighting, and spatial layout to evoke specific neural responses in viewers. By tapping into the brain's natural inclinations towards certain design features, architects can create environments that foster feelings of calmness, creativity, or even awe. Neuroaesthetics also plays a crucial role in shaping user experiences in virtual environments, where designers can manipulate sensory inputs to evoke desired emotional responses. This holistic approach to design considers not only the physical aspects of a space but also the psychological and emotional impact it has on individuals. The integration of neuroaesthetics into design and architecture has the potential to revolutionize the way we interact with our built environment. By designing spaces that are optimized for human well-being and cognitive function, we can create environments that enhance productivity, creativity, and overall quality of life. As our understanding of the brain continues to advance, the application of neuroaesthetics in design and architecture will only become more sophisticated, leading to the creation of truly immersive and transformative spaces that cater to the complex needs of the human mind.

# XI. NEUROTECHNOLOGY IN MILITARY AND DEFENSE

Neurotechnology has gained significant traction in military and defense sectors due to its potential to enhance soldiers' cognitive and physical capabilities on the battlefield. By utilizing BCIs, military personnel can communicate more efficiently, control advanced weaponry with their mind, and even receive real-time feedback on their physiological state. These advancements have the potential to revolutionize the way wars are fought, ensuring quicker decision-making processes and increased efficiency in combat situations. The integration of neurotechnology in the military also raises ethical concerns regarding the potential misuse of such technology for coercive purposes, highlighting the importance of ethical guidelines in its development and deployment. The application of neurotechnology in military and defense extends beyond the battlefield, with potential implications for intelligence gathering and strategic planning. BCIs could be used to enhance cognitive abilities, improve memory retention, and facilitate faster information processing, leading to more effective decision-making at all levels of military operations. The use of neurotechnology in training programs could help optimize skills acquisition and improve overall performance, resulting in a more agile and adaptive military force. These advancements have the potential to not only strengthen national security but also enhance the safety and well-being of military personnel in high-risk environments. As with any emerging technology, there are challenges and risks associated with the widespread adoption of neurotechnology in military and defense contexts. Concerns about data security, privacy, and the potential for unauthorized access to neural data are paramount, raising questions about the ethical implications of using BCIs in combat scenarios. The potential for adversaries to hack into BCIs and manipulate soldiers' thoughts or actions poses a significant threat to national security. As neurotechnology continues to advance, it is crucial for policymakers, researchers, and military leaders to address these challenges proactively and develop robust frameworks to ensure the responsible and ethical use of such technology in military applications.

## BCIs for Enhanced Soldier Performance

Advances in neurotechnology have paved the way for the development of BCIs aimed at enhancing soldier performance. By directly connecting the brain to external devices, BCIs have the potential to revolutionize how soldiers interact with technology on the battlefield. Through the use of implanted sensors or non-invasive devices, soldiers can communicate, control sophisticated equipment, and receive real-time feedback all through neural signals. This direct neural interface can significantly increase response times, accuracy, and overall tactical awareness, giving soldiers a decisive edge in complex and high-pressure situations. BCIs can enable soldiers to operate drones, vehicles, or other remote-controlled devices with unparalleled precision and efficiency. This technology has the potential to reduce cognitive load during missions, allowing soldiers to focus on critical decision-making tasks without being overwhelmed by the technical aspects of operating complex systems. By harnessing the power of the human brain, BCIs can enhance the cognitive capabilities and situational awareness of soldiers, ultimately improving mission outcomes and increasing overall effectiveness on the battlefield. BCIs offer the possibility of seamless integration with existing military systems and equipment, providing a customized and intuitive interface for soldiers to interact with their surroundings. This level of integration can streamline communication between soldiers, commanders, and support staff, creating a highly interconnected and responsive network on the battlefield. As BCIs continue to evolve and become more sophisticated, the potential for enhancing soldier performance becomes even more promising, ushering in a new era of human-machine collaboration in military operations.

## Neuroenhancement in Military Training

As neurotechnology continues to advance, the potential for its application in military training is becoming increasingly recognized. Neuroenhancement techniques, such as BCIs, hold promise for improving cognitive functions, enhancing learning and memory retention, and optimizing decision-making processes among military personnel. By integrating BCIs into training programs, soldiers can benefit from real-time feedback on their performance, enabling them to adjust their strategies and tactics more effectively. This could lead to improved situational awareness, quicker decision-making under pressure, and overall better performance in complex and high-stress environments. The use of neuroenhancement in military training could also have implications for the ethical considerations surrounding the enhancement of human capabilities. As BCIs have the potential to augment cognitive functions beyond their natural limits, questions arise about the fairness of providing such technologies to some individuals but not others. Concerns about the potential for misuse or unintended consequences of neuroenhancement in military contexts must be carefully considered. Ethical guidelines and regulations will need to be established to ensure that neuroenhancement technologies are used responsibly and ethically in the military setting. The integration of neuroenhancement techniques such as BCIs in military training has the potential to revolutionize the way soldiers are trained and equipped for their roles. By harnessing the power of BCIs to enhance cognitive functions and improve decision-making processes, military personnel could be better prepared to handle the challenges of modern warfare. Careful consideration must be given to the ethical implications and potential risks associated with the use of neuroenhancement technologies in military contexts to ensure that they are deployed responsibly and with the best interests of both the soldiers and society in mind.

## Ethical Considerations in Weaponized Neurotechnology

As the development of weaponized neurotechnology continues to progress, ethical considerations become increasingly crucial in ensuring the responsible use of these powerful tools. One key area of concern is the potential for misuse and manipulation of neural data obtained through BCIs for malicious purposes. The invasion of privacy through unauthorized access to individuals' thoughts and cognitive processes poses a significant threat, raising the need for stringent regulations to safeguard against such breaches. The risk of cognitive hacking, where individuals' brains are manipulated without their consent, introduces complex ethical dilemmas that must be carefully addressed to prevent exploitation and harm. Another ethical consideration in weaponized neurotechnology is the issue of consent and autonomy in the use of BCIs for military or defense purposes. The deployment of neurotechnologies in warfare raises questions about the ethical implications of using neural data to control or influence individuals' actions without their full understanding or consent. It is essential to establish clear guidelines and protocols for the ethical deployment of weaponized neurotechnology to ensure that individuals are fully informed and have agency over the use of their neural data in military contexts. Respecting individuals' autonomy and rights to privacy is paramount in shaping ethical practices surrounding the integration of BCIs into military operations. The potential long-term societal impacts of weaponized neurotechnology necessitate a broader ethical discussion surrounding its development and deployment. The normalization of using BCIs for military or security purposes may have far-reaching consequences on individuals' perceptions of privacy, autonomy, and agency over their thoughts and actions. Addressing these ethical considerations requires a comprehensive approach that considers not only the immediate risks and benefits of weaponized neurotechnology but also the broader ethical implications for society as a whole. As researchers and policymakers navigate the complex terrain of weaponized neurotechnology, prioritizing ethical considerations is essential to ensure the responsible and ethical development of these powerful tools.

# XII. NEUROTECHNOLOGY AND NEUROPLASTICITY

Neuroplasticity plays a vital role in the development and success of BCIs. As the brain's ability to reorganize itself and form new neural connections in response to learning or experience, neuroplasticity is key to adapting to the use of BCI technology. By harnessing the brain's innate plasticity, researchers can optimize the functioning of BCIs and improve user performance over time. This dynamic interplay between neuroplasticity and neurotechnology highlights the importance of ongoing research in understanding how the brain can adapt to new technologies. Neuroplasticity can also impact the design and implementation of BCIs, as it allows for the customization of interfaces based on individual neural responses. Through targeted training and feedback mechanisms, BCIs can capitalize on neuroplastic changes to enhance user experiences and outcomes. By tailoring BCI systems to leverage the brain's plastic nature, researchers can optimize performance, increase efficiency, and improve overall usability for individuals using these technologies. This personalized approach underscores the potential for BCIs to revolutionize human-computer interaction and foster greater integration of technology into daily life. In essence, the symbiotic relationship between neuroplasticity and neurotechnology holds great promise for the future of BCIs. By understanding and leveraging the brain's adaptive capabilities, researchers can enhance the effectiveness and feasibility of these innovative interfaces. As neuroplasticity continues to be a driving force in shaping the development and application of BCIs, the field stands to benefit from ongoing advancements in both neuroscience and technology. By embracing the principles of neuroplasticity, BCIs can pave the way for a future where seamless interaction between humans and machines is not only achievable but also transformative in its potential impact on society.

## Harnessing Neuroplasticity for Cognitive Enhancement

Research in the field of neuroscience has shown that the brain possesses a remarkable ability to reorganize itself and adapt to new challenges, a phenomenon known as neuroplasticity. Harnessing neuroplasticity for cognitive enhancement has become a promising area of research, with significant implications for improving various aspects of human cognition, such as memory, attention, and learning. By understanding the mechanisms underlying neuroplasticity, researchers are exploring ways to leverage this natural ability of the brain to enhance cognitive functions through innovative interventions and technologies. One approach to harnessing neuroplasticity for cognitive enhancement is the development of BCIs. BCIs are devices that establish direct communication between the brain and external devices, allowing individuals to control computers, prosthetic limbs, or other gadgets using their thoughts alone. By pairing neurofeedback techniques with advanced signal processing algorithms, BCIs can facilitate targeted brain activation patterns to enhance cognitive abilities. This offers a personalized and adaptive approach to cognitive enhancement that can cater to individual needs and challenges, ultimately leading to more effective interventions. The integration of neuroplasticity and BCIs opens up new possibilities for enhancing cognitive functions in both healthy individuals and those with neurological conditions. From improving memory and attention in healthy adults to restoring motor functions in individuals with neurodegenerative diseases, the applications of harnessing neuroplasticity for cognitive enhancement are vast and diverse. As researchers continue to unravel the complexities of brain plasticity and develop sophisticated BCI technologies, the future holds great promise for unlocking the full potential of the human brain and revolutionizing the way we interact with technology for cognitive enhancement.

## Neurofeedback Training for Brain Plasticity

Neurofeedback training has emerged as a promising technique to enhance brain plasticity and optimize cognitive function. By providing real-time feedback on brain activity, individuals can learn to self-regulate their neural patterns, leading to improved connectivity, neural efficiency, and overall brain health. This form of training enables individuals to harness the brain's inherent ability to adapt and reorganize itself in response to new experiences, ultimately promoting neuroplasticity in targeted brain regions. Through consistent practice and reinforcement, neurofeedback has the potential to facilitate lasting changes in brain structure and function, paving the way for enhanced cognitive performance and mental well-being. Research has shown that neurofeedback training can be applied to a wide range of clinical conditions, such as ADHD, depression, anxiety, and post-traumatic stress disorder. By targeting specific brain regions associated with these conditions, neurofeedback aims to modify dysfunctional neural patterns and restore optimal brain function. Studies have demonstrated the efficacy of neurofeedback in improving symptoms and reducing the need for medication in various populations. The personalized nature of neurofeedback allows for individualized treatment plans tailored to each person's unique brain activity, offering a non-invasive and drug-free approach to mental health care. The integration of neurofeedback training into mainstream medical practice holds great promise for improving patient outcomes and revolutionizing the field of neuromodulation. As neurotechnology continues to advance, the development of more accessible and user-friendly neurofeedback systems will democratize access to this cutting-edge therapy. By harnessing the power of the brain's plasticity through targeted training protocols, neurofeedback has the potential to unlock new pathways for cognitive enhancement, brain rehabilitation, and mental health treatment. Neurofeedback training for brain plasticity represents a transformative approach to optimizing brain function and unlocking the full potential of the human mind.

## Implications for Learning and Skill Acquisition

As we delve deeper into the world of neurotechnology and BCIs, the implications for learning and skill acquisition become increasingly apparent. One of the most fascinating aspects of BCIs is their potential to revolutionize the way we interact with technology, particularly in educational settings. By allowing direct communication between the brain and external devices, BCIs have the power to enhance the learning process, making it more immersive, intuitive, and personalized. Students could potentially learn new languages or complex subjects more efficiently by directly interfacing with educational software using their thoughts, bypassing traditional input methods like typing or clicking. The ability to directly access and manipulate information through BCIs can also have profound implications for skill acquisition. Imagine a scenario where individuals can quickly learn new skills or tasks by directly downloading relevant knowledge into their brains. This could significantly reduce the time and effort required to acquire new abilities, opening up a world of possibilities for personal and professional development. BCIs could aid in skill retention and transfer by facilitating the seamless integration of new information into existing knowledge networks, leading to more robust and adaptable skill sets. The integration of BCIs into learning environments has the potential to transform traditional education systems, making them more efficient, engaging, and effective. By harnessing the power of neurotechnology, individuals could unlock new learning pathways and enhance their skill acquisition abilities in ways previously unimaginable. It is crucial to consider the ethical and societal implications of such advancements, ensuring that the benefits of BCIs are balanced with responsible implementation and consideration for the broader impact on individuals and society as a whole.

# XIII. NEUROTECHNOLOGY IN NEUROSCIENCE RESEARCH

As neurotechnology continues to advance, it holds great promise for revolutionizing neuroscience research. BCIs have opened up new avenues for understanding the complexities of the human brain and its functions. By allowing direct communication between the brain and external devices, researchers can explore neural activity and cognitive processes in ways previously unimaginable. This technology has the potential to uncover fundamental insights into brain functioning and could lead to breakthroughs in treating neurological disorders and enhancing human capabilities. One key aspect of neurotechnology in neuroscience research is its ability to bridge the gap between basic research and clinical applications. By utilizing BCIs, researchers can not only study brain activity in real-time but also develop novel interventions for individuals with motor disabilities or neurodegenerative conditions. This translational aspect of neurotechnology demonstrates its potential to directly impact patient care and improve quality of life. The ability to decode neural signals and manipulate them opens up new possibilities for both understanding the brain and developing innovative treatment strategies. The integration of neurotechnology in neuroscience research has the potential to shape the future of human-machine interactions. As BCIs become more sophisticated and accessible, they could fundamentally transform the way we interact with technology, from controlling devices with our thoughts to enhancing our cognitive abilities. The implications of this technology extend beyond the realm of research, offering glimpses into a future where humans and machines collaborate seamlessly. By embracing the opportunities presented by neurotechnology, we stand to unlock the full potential of the human brain and pave the way for a new era of scientific discovery and technological innovation.

## Advancements in Brain Mapping and Connectivity Studies

Recent advancements in brain mapping and connectivity studies have revolutionized our understanding of the human brain. With the development of sophisticated imaging techniques such as fMRI and EEG, researchers can now visualize neural activity in unprecedented detail. These tools allow for the mapping of brain regions involved in various cognitive functions, providing valuable insights into how different areas of the brain communicate and work together to process information. By studying the connectivity patterns within the brain, scientists can uncover the underlying neural networks that govern our behavior, emotions, and thoughts. Advancements in brain mapping have led to breakthroughs in the field of neurotechnology, particularly in the development of BCIs. BCIs enable direct communication between the brain and external devices, opening up a world of possibilities for individuals with motor disabilities or neurological conditions. By decoding neural signals and translating them into commands, BCIs can empower users to control prosthetic limbs, computer applications, or even communicate through thought alone. These devices not only enhance the quality of life for those with impairments but also hold promise for enhancing human capabilities beyond what is naturally possible. As brain mapping technologies continue to progress, researchers are uncovering the intricate web of connections that underlie our cognitive functions and behaviors. By elucidating the complex pathways through which information flows in the brain, scientists are gaining new insights into the mechanisms of learning, memory, and decision-making. This deeper understanding of neural connectivity not only sheds light on the underlying causes of neurological disorders but also provides a foundation for developing targeted interventions and treatments. The advancements in brain mapping and connectivity studies are paving the way for a future where we can harness the full potential of the human brain to enhance our lives and unlock new possibilities for humanity.

## Neuroimaging Techniques for Understanding Brain Function

Neuroimaging techniques play a crucial role in understanding the complex functions of the brain. These techniques, such as fMRI and EEG, provide researchers with valuable insights into brain activity in real-time. By measuring blood flow or electrical activity in different brain regions, neuroimaging allows for the mapping of cognitive processes, sensory perception, motor control, and emotional responses. Through the analysis of brain images, researchers can identify neural networks, patterns of activation, and abnormalities associated with various neurological and psychiatric disorders. Advances in neuroimaging technology have led to a deeper understanding of the brain's intricate workings. DTI enables researchers to investigate the structural connectivity of the brain by tracing the pathways of white matter fibers. This has important implications for studying neural development, neuroplasticity, and the effects of brain injuries. PET scans can detect changes in neurotransmitter levels, providing valuable information about conditions such as Alzheimer's disease, Parkinson's disease, and schizophrenia. The integration of neuroimaging with other fields such as AI and machine learning has opened up new possibilities for brain research. By analyzing large datasets of brain images, researchers can develop predictive models for diagnosing neurological disorders, monitoring disease progression, and evaluating treatment outcomes. The combination of advanced neuroimaging techniques with computational tools has the potential to revolutionize personalized medicine and improve patient care by enabling precise targeting of interventions based on individual brain characteristics. In essence, neuroimaging techniques continue to advance our understanding of the brain and pave the way for innovative approaches to studying cognition, behavior, and brain function.

## Neurotechnology in Studying Neurological Disorders

Neurotechnology has revolutionized the study and treatment of neurological disorders by offering innovative solutions that were previously unimaginable. The development of BCIs has paved the way for a deeper understanding of the brain's inner workings and its potential to overcome various neurological challenges. By bridging the gap between neuroscience and technology, BCIs have opened new avenues for exploring the complexities of the human brain and its applications in the field of medicine. Through the use of advanced sensors, signal processing, and decoding algorithms, BCIs have enabled researchers to harness the power of neural signals to enhance diagnosis, treatment, and rehabilitation of neurological conditions. One of the key advantages of neurotechnology in studying neurological disorders is its ability to provide real-time feedback and personalized interventions based on an individual's neural activity. This personalized approach allows for targeted therapies that can adapt to the unique needs of each patient, improving treatment outcomes and quality of life. BCIs have shown promising results in restoring motor functions for individuals with disabilities, such as controlling prosthetic limbs through thought alone. By interfacing with the brain directly, BCIs offer a new level of precision and control that was previously unattainable with traditional medical interventions. Despite the incredible advancements in neurotechnology, there remain challenges and ethical considerations that must be addressed to ensure the responsible development and deployment of BCIs. Privacy concerns, potential risks of invasive procedures, and the implications of manipulating neural data are all important factors that need to be carefully considered. The societal impact of widespread adoption of BCIs must be carefully evaluated to ensure equitable access and prevent further disparities in healthcare. By addressing these challenges thoughtfully, neurotechnology has the potential to revolutionize the field of neuroscience and lead to groundbreaking discoveries in the study and treatment of neurological disorders.

# XIV. NEUROTECHNOLOGY AND HUMAN RIGHTS

Neurotechnology presents a myriad of possibilities for enhancing human capabilities and revolutionizing the way we interact with technology. As this field continues to advance, it brings to the forefront important ethical considerations that must be addressed to protect human rights. One of the key concerns regarding neurotechnology is the potential for invasive methods, such as implanting devices into the brain, to infringe upon individuals' right to privacy. The collection and storage of neural data raise questions about who has access to this information and how it may be used, highlighting the need for robust regulations to safeguard personal data and prevent misuse. The development of neurotechnology brings into question the autonomy and agency of individuals, particularly concerning the potential for direct manipulation of the human brain. This raises concerns about the implications for personal freedom and the ability to control one's own thoughts and actions. As neurotechnology advances, there is a risk of creating a society where individuals' thoughts and emotions can be monitored and manipulated, leading to a potential erosion of civil liberties and human rights. Thus, it is crucial to establish clear guidelines and ethical frameworks to ensure that the use of neurotechnology respects and upholds the rights and dignity of individuals. In light of these ethical challenges, it is essential for policymakers, researchers, and technology developers to engage in meaningful dialogue to address the complex issues at the intersection of neurotechnology and human rights. By fostering interdisciplinary collaboration and promoting transparency in the development and deployment of neurotechnologies, we can work towards harnessing the transformative potential of these advancements while upholding fundamental human rights. Navigating the ethical considerations surrounding neurotechnology is crucial to ensuring that these innovations contribute to enhancing human well-being and fostering a more equitable and rights-respecting society.

## Access to Neurotechnological Advancements as a Human Right

Neurotechnological advancements have the potential to revolutionize the way humans interact with technology, offering new possibilities for communication, entertainment, and medical treatment. As these technologies continue to evolve, the question of access to neurotechnological advancements emerges as a crucial issue of human rights. Ensuring equal access to these innovations is essential to prevent creating or exacerbating societal disparities based on technological privilege. The right to access neurotechnological advancements can be seen as a fundamental human right, enabling individuals to reach their full potential, participate fully in society, and benefit from the latest advancements in science and technology. Individuals with disabilities stand to benefit significantly from the advancements in neurotechnology, particularly through the development of BCIs that can enhance their quality of life and facilitate communication and mobility. By granting these individuals access to neurotechnological advancements, society can promote inclusivity and support their right to participate fully in social, economic, and cultural life. Access to BCIs can empower individuals with disabilities to overcome physical limitations and lead more independent and fulfilling lives, aligning with the principles of social justice and equality. In addition to improving the quality of life for individuals with disabilities, ensuring access to neurotechnological advancements as a human right can also drive innovation and progress in various fields. By democratizing access to BCIs and other neurotechnological tools, we can tap into the diverse perspectives and talents of a wider range of individuals, fostering creativity and pushing the boundaries of what is possible. Recognizing access to neurotechnological advancements as a human right is not only a matter of social justice but also a strategic decision that can lead to a more inclusive, innovative, and prosperous society for all.

## Ethical Considerations in Cognitive Liberty

Ethical considerations play a crucial role in the field of cognitive liberty, particularly concerning the development and implementation of BCIs. As these technologies advance, questions regarding individual autonomy, privacy, and consent become more salient. The ability to access and manipulate neural data raises concerns about the potential for infringement on cognitive privacy, as well as the risk of unauthorized access to individuals' innermost thoughts and feelings. The prospect of direct brain manipulation through BCIs raises ethical dilemmas related to personal identity and agency, as well as the potential for coercion or manipulation by external forces. Issues of equity and social justice come into play when considering the deployment of BCIs in various contexts. There is a risk that these technologies may exacerbate existing disparities, as access to cutting-edge neurotechnologies may be limited to certain privileged groups.

The use of BCIs in areas such as security and law enforcement raises concerns about the potential for discrimination and bias in decision-making processes based on neural data. This underscores the importance of ensuring that regulations and ethical guidelines are in place to govern the ethical use of BCIs and protect vulnerable populations from harm. In navigating the ethical landscape of cognitive liberty and BCIs, it is essential to strike a balance between promoting technological innovation and safeguarding individual rights and freedoms. Stakeholders in the development and deployment of BCIs must actively engage in ethical reflection and discourse to address potential risks and challenges. By fostering a culture of responsible innovation and ethical governance, we can harness the transformative potential of neurotechnologies while upholding fundamental principles of autonomy, privacy, and justice in the digital age. A mindful approach to ethical considerations in cognitive liberty can ensure that the benefits of BCIs are maximized for the betterment of society as a whole.

## Neurotechnology and the Right to Mental Privacy

One of the most pressing concerns surrounding the development and widespread use of neurotechnology, particularly BCIs, is the issue of mental privacy. As these devices become more sophisticated and capable of decoding complex brain signals, the potential for invasion of individuals' innermost thoughts and emotions is a very real possibility. The right to mental privacy, therefore, is a fundamental consideration that must be addressed as we continue to integrate these technologies into various aspects of our lives. Without robust safeguards in place to protect the sanctity of our mental processes, there is a risk of exploitation and manipulation that could have far-reaching implications. In order to uphold the right to mental privacy in the age of neurotechnology, it is essential to establish clear ethical guidelines and legal frameworks that govern the collection, use, and storage of neural data. This includes ensuring that individuals have full control over their own brain data and that it cannot be accessed or utilized without their explicit consent. Measures must be put in place to prevent unauthorized access or hacking of neural interfaces, which could lead to breaches of personal privacy and the misuse of sensitive information. By prioritizing the protection of mental privacy in the design and implementation of BCIs, we can mitigate the potential risks associated with these technologies and foster a more ethical and responsible approach to their development and use. Ongoing dialogue and collaboration between researchers, policymakers, ethicists, and the general public are crucial in shaping the ethical discourse surrounding neurotechnology and mental privacy. By engaging in open and transparent discussions about the implications of BCIs on privacy and autonomy, we can work towards consensus on best practices and guidelines that protect individuals' mental privacy while still allowing for the advancement and innovation of these technologies. The right to mental privacy is a foundational human right that must be upheld and safeguarded in the face of rapid technological advancements, ensuring that individuals retain control over their most intimate thoughts and experiences in an increasingly interconnected world.

# XV. NEUROTECHNOLOGY IN SPORTS PERFORMANCE

Neurotechnology has started to play a significant role in enhancing sports performance, offering athletes innovative ways to train, track progress, and optimize their capabilities. By utilizing BCIs, athletes can access real-time feedback on their cognitive and physiological states during training or competition. This technology allows for personalized training programs based on individual brain responses, leading to more efficient and effective performance improvements. Neurotechnology can help athletes enhance their mental focus, reduce stress, and even control aspects of their physical performance through mental commands. BCI technology in sports performance has the potential to revolutionize the way athletes train and compete by providing a deeper understanding of brain activity and its implications on physical performance. Through the use of EEG sensors and advanced data analysis, athletes can optimize their training regimens, improve decision-making processes, and enhance overall performance outcomes. The integration of neurotechnology in sports settings opens up new possibilities for athletes to achieve peak performance levels while reducing the risk of injury and improving recovery times. By tapping into the power of the brain-body connection, athletes can unlock their full potential and push the boundaries of human performance in sports. As neurotechnology continues to advance, the ethical implications of its use in sports performance must be carefully considered. Issues such as privacy, data security, and potential advantages for elite athletes over others need to be addressed to ensure fair competition and ethical practices in sports. The potential for enhancing performance beyond natural capabilities raises questions about the line between enhancement and unfair advantage. By establishing clear guidelines and ethical frameworks for the incorporation of neurotechnology in sports, we can harness its benefits while safeguarding the integrity of sport and promoting a level playing field for all athletes.

## BCIs for Athlete Training and Monitoring

BCIs have shown great promise in the realm of athlete training and monitoring, offering unique opportunities for enhancing performance and gaining valuable insights into the physiological and cognitive aspects of sports. These interfaces enable real-time monitoring of brain activity, allowing coaches and trainers to assess athletes' mental states, focus levels, and cognitive processes during training or competition. By analyzing this data, adjustments can be made to training regimens, strategies, and even equipment to optimize performance and minimize the risk of injuries. BCIs can be used to create immersive virtual environments that simulate competitive scenarios, providing athletes with a safe space to practice and refine their skills in a controlled setting. In addition to training, BCIs can also play a crucial role in monitoring athletes' physiological responses and overall well-being during strenuous activities. By tracking brain signals, heart rate, oxygen levels, and other relevant biomarkers, BCIs can provide real-time feedback on an athlete's physical condition, helping to prevent overexertion, dehydration, and other potential health risks. This data can also be used to tailor recovery protocols and nutrition plans based on individual needs and responses. BCIs can assist in the early detection of concussions or other traumatic brain injuries, enabling prompt medical intervention and ensuring the safety and well-being of athletes. The integration of BCIs in athlete training and monitoring represents a significant advancement in sports science and performance optimization. By leveraging the power of neurotechnology, coaches, trainers, and athletes can gain a deeper understanding of the mind-body connection and unlock new ways to improve performance, prevent injuries, and enhance overall well-being. As this technology continues to evolve, the potential for BCIs to revolutionize the sports industry and elevate the capabilities of athletes will only grow, paving the way for a future where human potential is maximized through the seamless integration of technology and physiology.

## Cognitive Enhancement in Sports Psychology

Cognitive enhancement in sports psychology has become a focal point for researchers and athletes alike, seeking to maximize performance and gain a competitive edge. By utilizing neurotechnology such as BCIs, athletes can tap into their cognitive abilities in ways previously thought impossible. These advancements hold the potential to revolutionize the way athletes train and compete, opening up new possibilities for optimizing mental processes such as focus, decision-making, and response time. Incorporating BCI technology into sports psychology not only enhances individual athletic performance but also contributes to the overall understanding of human cognition in high-pressure situations. The integration of cognitive enhancement techniques in sports psychology through neurotechnology represents a paradigm shift in how athletes approach mental training. With the ability to monitor and modulate brain activity in real-time, athletes can fine-tune their cognitive functions to achieve peak performance levels. This personalized approach to mental conditioning allows for targeted interventions based on individual neurocognitive profiles, enhancing the effectiveness of training regimens and performance strategies. By harnessing the power of BCIs in sports psychology, athletes can unlock their full cognitive potential and push the boundaries of human athletic achievement. The implications of cognitive enhancement in sports psychology extend beyond individual performance gains to broader societal considerations. As neurotechnology continues to advance, ethical dilemmas regarding fairness, consent, and the potential for enhancement beyond natural capabilities come to the forefront. Addressing these complex issues will be paramount in ensuring the responsible and ethical use of cognitive enhancement tools in sports psychology. By navigating these challenges thoughtfully and ethically, the integration of BCIs in sports psychology has the potential to not only enhance athletic performance but also spark meaningful discussions about the future of human enhancement and the boundaries of cognitive capabilities.

## Ethical Implications of Performance-Enhancing Neurotechnology

One of the most pressing ethical considerations surrounding the use of performance-enhancing neurotechnology is the potential for creating unequal playing fields in various domains, such as academics, sports, and the workplace. As individuals with access to these technologies may gain cognitive or physical advantages over their peers, questions arise regarding fairness and the impact on competition. In sports, athletes using neurotechnology to enhance their cognitive abilities or physical performance could have an unfair advantage over those who do not have access to such enhancements. This raises concerns about the integrity of competitions and the implications for those who may feel pressured to use these technologies to remain competitive. The use of performance-enhancing neurotechnology can blur the lines between natural abilities and artificially augmented capabilities, raising questions about authenticity and the concept of human enhancement. As individuals enhance their cognitive functions or physical skills through neurotechnology, the boundary between what is considered "natural" and "enhanced" becomes increasingly unclear. This challenges traditional notions of human capabilities and raises philosophical questions about the nature of identity and authenticity. The potential for individuals to alter their personalities or cognitive processes through neurotechnology raises concerns about autonomy and free will, as individuals may be influenced or manipulated in ways that could have far-reaching consequences. Another key ethical consideration in the use of performance-enhancing neurotechnology is the potential impact on societal norms and values. As these technologies become more widespread and accessible, they have the potential to shift societal expectations regarding human capabilities and performance. This can lead to increased pressure on individuals to enhance themselves through neurotechnology in order to meet societal standards or expectations. The normalization of using neurotechnology to enhance performance may exacerbate existing inequalities, as those who cannot afford or access these technologies may be left at a disadvantage. This highlights the importance of considering the broader societal implications of using performance-enhancing neurotechnology and ensuring that ethical guidelines are in place to protect individuals and promote fairness.

# XVI. NEUROTECHNOLOGY AND ENVIRONMENTAL SUSTAINABILITY

Neurotechnology, particularly the development of BCIs, holds immense potential in driving environmental sustainability initiatives. By leveraging the power of BCI technology, individuals can enhance their cognitive abilities, thereby enabling more effective decision-making processes in relation to sustainability practices. BCI could facilitate real-time monitoring of environmental data, allowing for more accurate and immediate responses to environmental challenges such as climate change or conservation efforts. This increased awareness and data-driven approach can lead to more efficient resource management and informed policy interventions to protect the planet. The integration of BCI with environmental sensors and devices can enable personalized sustainability solutions tailored to individual preferences and behaviors. By providing users with direct feedback on their environmental impact, BCI can promote sustainable practices at the individual level, fostering a culture of conservation and eco-friendly habits. By harnessing the cognitive abilities of individuals through BCI, innovative solutions for sustainable living can be developed, such as smart home technologies that optimize energy usage based on the user's cognitive input and preferences. The fusion of neurotechnology and environmental sustainability has the potential to revolutionize the way humans interact with and protect the environment. By harnessing the power of BCI, individuals can become more actively engaged in sustainability efforts, leading to a more conscious and environmentally responsible society. As BCI technology continues to advance, it is crucial to explore its applications in the realm of environmental sustainability to leverage its transformative capabilities for the betterment of the planet and future generations.

## Applications of BCI in Environmental Monitoring

As technology continues to advance, the applications of BCIs in environmental monitoring have started to gain traction. By integrating BCI technology with environmental sensors, researchers can develop innovative ways to track and analyze environmental data in real-time. This can lead to more efficient and accurate monitoring of various environmental factors such as air quality, water pollution, and biodiversity. With the ability to interpret brain signals, BCIs can provide researchers with a unique perspective on how individuals interact with their environment, enabling a deeper understanding of human-environment interactions. One of the key benefits of using BCIs for environmental monitoring is the potential to enhance environmental conservation efforts. By leveraging the power of neural signals, researchers can gather valuable insight into how individuals perceive and respond to environmental changes. This information can help inform conservation strategies and policies, leading to more effective and sustainable environmental management practices. BCI technology can enable early detection of environmental issues, allowing for timely intervention and mitigation measures to be implemented. The use of BCIs in environmental monitoring can pave the way for more personalized and responsive environmental solutions. By analyzing individual brain signals in relation to environmental stimuli, researchers can tailor interventions to meet specific needs and preferences. This level of personalization can lead to greater engagement and participation in environmental conservation efforts, ultimately leading to a more environmentally conscious society. The integration of BCI technology in environmental monitoring holds great promise for enhancing our understanding of the environment and driving sustainable conservation actions.

## Neurofeedback for Sustainable Behavior Change

Neurofeedback has shown promise as a tool for sustainable behavior change by enabling individuals to self-regulate their brain activity. By providing real-time feedback on brainwave patterns, neurofeedback allows users to modify their thoughts, emotions, and behaviors consciously. This self-regulation mechanism can lead to long-lasting changes in behavior, as individuals learn to control their brain activity in response to specific cues or stimuli. Through repeated sessions of neurofeedback training, individuals can strengthen neural pathways associated with desired behaviors, ultimately leading to more sustainable changes in habits and actions. This process of neuroplasticity lays the foundation for behavior change by rewiring the brain to support positive habits and decision-making. Neurofeedback has the potential to address underlying neural mechanisms that contribute to maladaptive behaviors, such as addiction or impulsivity. By targeting specific brain regions or networks involved in these behaviors, neurofeedback can help individuals develop more adaptive patterns of neural activity. Studies have shown that neurofeedback training can modulate activity in the prefrontal cortex, a region crucial for executive functions like self-control and decision-making. By enhancing the functioning of this brain region through neurofeedback, individuals may experience improved self-regulation and reduced impulsivity, promoting sustainable behavior change over time. The personalized nature of neurofeedback makes it a powerful tool for sustainable behavior change, as it tailors interventions to the individual's unique brain activity patterns. By capturing real-time data on brainwave activity and providing immediate feedback, neurofeedback can adapt its training protocols to optimize outcomes for each user. This individualized approach allows for targeted interventions that address specific neural mechanisms underlying problematic behaviors, increasing the efficacy and sustainability of behavior change efforts. As neurofeedback technology continues to advance, integrating machine learning algorithms and adaptive protocols, the potential for personalized, sustainable behavior change through neurofeedback training is only expected to grow.

## Neurotechnology in Promoting Eco-Friendly Practices

Neurotechnology has the potential to revolutionize eco-friendly practices by enhancing human interactions with the environment through BCIs. By leveraging the power of neurotechnology, individuals can be more attuned to their surroundings and make more environmentally conscious decisions. BCIs can be used to monitor brain activity in response to different environmental stimuli, providing valuable insights into how individuals react to their surroundings. This data can then be used to develop strategies to promote eco-friendly behaviors, such as reducing energy consumption or waste production. Through the combination of neuroscience and technology, neurotechnology can play a key role in driving sustainable practices and promoting a more environmentally friendly way of life. Neurotechnology can be utilized to create innovative solutions for environmental challenges by tapping into the cognitive abilities of individuals. By incorporating BCI technology into environmental monitoring systems, for instance, real-time feedback on environmental conditions can be provided based on brain signals. This can enable more effective and efficient resource management, leading to better conservation efforts and reduced environmental impact. Neurotechnology can enhance communication and collaboration among stakeholders in the environmental sector, facilitating the exchange of ideas and strategies for sustainable development. By harnessing the potential of neurotechnology, eco-friendly practices can be promoted and implemented on a broader scale, fostering a more environmentally conscious society. The integration of neurotechnology in promoting eco-friendly practices holds immense promise for shaping a sustainable future. By leveraging the capabilities of BCIs, individuals can become more engaged and proactive in addressing environmental challenges. Through innovative applications of neurotechnology, such as monitoring brain activity in response to environmental stimuli and enhancing communication in the environmental sector, sustainable practices can be more effectively promoted and implemented. As we continue to explore the possibilities of neurotechnology, we are paving the way for a more environmentally conscious society, where technology and neuroscience converge to create a healthier and more sustainable world for future generations.

# XVII. NEUROTECHNOLOGY AND AGING POPULATION

As the aging population continues to grow globally, the intersection of neurotechnology and elderly care is becoming increasingly crucial. With advancements in BCIs, there is a potential to revolutionize the way we address the cognitive and neurological challenges associated with aging. By harnessing the power of neural signals, BCIs offer a promising avenue for enhancing cognitive functions, improving communication, and restoring independence in older adults. One key area where neurotechnology can have a significant impact on the aging population is in the realm of neurorehabilitation. As individuals age, they may experience a decline in motor skills or cognitive function due to neurodegenerative diseases or age-related conditions. BCIs can be utilized to facilitate neuroplasticity, enabling older adults to regain lost functions through targeted neural stimulation and cognitive training. This can lead to improved quality of life, increased autonomy, and enhanced emotional well-being for the elderly. The integration of BCIs in elderly care can also help address social isolation and loneliness among older adults. By enabling seamless communication through neural interfaces, individuals can stay connected with their loved ones, engage in activities, and participate in social interactions that contribute to their overall mental and emotional health. As we continue to explore the potential of neurotechnology in aging populations, it is essential to consider the ethical implications, privacy concerns, and accessibility issues to ensure that these innovations benefit older adults in a safe and equitable manner.

## Cognitive Support for Elderly Individuals

In addressing the cognitive support needs of elderly individuals, it is crucial to consider the specific challenges this population faces as they age. Cognitive decline is a common issue among the elderly, affecting memory, decision-making, and overall cognitive functioning. Providing cognitive support through interventions such as brain training exercises, cognitive therapy, and personalized technology solutions can help mitigate the effects of age-related cognitive decline. By engaging in regular cognitive activities and utilizing technology that supports cognitive functions, elderly individuals can enhance their mental acuity and overall quality of life. One effective approach to cognitive support for elderly individuals is the use of personalized technology solutions that cater to their specific cognitive needs. BCIs can be particularly beneficial in this regard, as they offer a direct link between the brain and external devices, allowing for real-time monitoring and feedback. By incorporating BCIs into cognitive support programs for the elderly, individuals can engage in interactive and personalized cognitive exercises that target their unique cognitive challenges. This personalized approach can lead to more effective cognitive support outcomes and improve overall cognitive functioning in elderly individuals. The integration of BCIs in cognitive support programs for the elderly can also enhance social interactions and engagement, which are crucial aspects of healthy aging. By utilizing BCIs for communication and cognitive engagement activities, elderly individuals can stay connected with their peers, families, and communities, reducing feelings of isolation and loneliness. This social aspect of cognitive support is essential for overall well-being and mental health in elderly individuals, highlighting the potential of BCIs to not only improve cognitive functioning but also enhance quality of life and social connectedness in the elderly population.

## Neurorehabilitation for Age-Related Cognitive Decline

In the field of neurorehabilitation for age-related cognitive decline, BCIs have shown promising potential in enhancing cognitive function and quality of life for older adults. By utilizing neurofeedback training and cognitive exercises tailored to individual needs, BCIs can target specific cognitive domains such as memory, attention, and executive function. These personalized interventions can help strengthen neural connections and improve cognitive resilience against age-related decline. BCIs can provide real-time feedback on brain activity, enabling individuals to actively engage in their cognitive rehabilitation and track their progress over time. BCIs can be integrated with other neurorehabilitation techniques such as transcranial magnetic stimulation (TMS) and neurofeedback to create a comprehensive and synergistic approach to cognitive training. This combination of interventions leverages the power of neuromodulation and neural plasticity to promote recovery and enhance cognitive functioning in older adults. By stimulating specific brain regions involved in memory consolidation, attention modulation, and cognitive control, BCIs can facilitate neuroplasticity and support the reorganization of neural networks to compensate for age-related changes in the brain. The non-invasive nature of BCIs makes them safe and accessible for older adults, allowing for home-based cognitive training and monitoring. With the advancements in wearable BCI technology, older adults can benefit from continuous cognitive support and feedback in their daily lives, promoting long-term cognitive health and well-being. As research in neurorehabilitation continues to evolve, BCIs hold great promise in revolutionizing the field and improving the cognitive outcomes of aging individuals, ultimately enhancing their quality of life and independence.

## Ethical Considerations in Enhancing Quality of Life for Seniors

Ethical considerations play a crucial role in the development and implementation of technologies aimed at enhancing the quality of life for seniors. As advancements in neurotechnology, such as BCIs, continue to progress, ethical dilemmas regarding autonomy, privacy, and consent become increasingly relevant. Seniors may face challenges related to decision-making capacity, making it essential to ensure that any enhancements to their quality of life are in line with their values and preferences. Issues surrounding data security and informed consent must be carefully navigated to protect vulnerable populations from potential harm. In the context of seniors, ethical considerations extend beyond individual autonomy to encompass broader societal implications. The use of BCIs in older populations raises questions about resource allocation, access to care, and equitable distribution of technological advancements. As such, policymakers and healthcare providers must grapple with the ethical complexities of balancing the potential benefits of neurotechnology for seniors with the need to address disparities in healthcare access and affordability. Issues of social justice and inclusivity must be carefully considered to ensure that advancements in quality of life do not exacerbate existing inequalities among seniors. Addressing ethical considerations in enhancing the quality of life for seniors through neurotechnology requires a nuanced approach that prioritizes respect for individual autonomy, dignity, and well-being. By engaging in transparent dialogue with seniors, caregivers, and other stakeholders, ethical frameworks can be developed to guide the responsible and equitable implementation of BCIs and other neurotechnologies. In doing so, we can leverage the transformative potential of these technologies to not only improve the quality of life for seniors but also promote social justice and human flourishing across the lifespan.

# XVIII. NEUROTECHNOLOGY AND GLOBAL HEALTH

Another important aspect of neurotechnology is its potential impact on global health. BCIs have the capability to revolutionize healthcare by providing innovative solutions for a wide range of medical conditions. BCIs can be used in the treatment of motor disabilities, allowing individuals with paralysis to control prosthetic limbs or communicate through speech synthesis. This can significantly improve the quality of life for patients with conditions such as spinal cord injuries or ALS, offering them greater independence and autonomy. Neurotechnology can also play a crucial role in neurorehabilitation for stroke patients. By using BCIs to facilitate brain plasticity and enhance neural connections, individuals recovering from a stroke can experience improved motor function and cognitive abilities. This personalized approach to rehabilitation shows promising results in restoring lost functions and promoting recovery. The potential of BCIs to treat neurological and psychiatric disorders, such as epilepsy or depression, opens up new avenues for targeted interventions and personalized therapies, ultimately improving outcomes for patients worldwide. The intersection of neurotechnology and global health holds immense promise for the future of healthcare. As BCIs continue to evolve and become more sophisticated, the potential applications in medical settings are vast. From enhancing mobility for individuals with disabilities to facilitating neurorehabilitation after neurological injuries, the transformative impact of neurotechnology on global health cannot be understated. By addressing ethical and societal challenges, and leveraging the power of BCIs for medical advancements, we can strive towards a future where neurotechnology plays a central role in improving health outcomes and enhancing quality of life for individuals around the world.

## BCI Applications in Developing Countries

In the context of developing countries, the potential applications of BCIs are vast and could significantly impact various sectors such as healthcare, education, and communication. One crucial area where BCIs could make a difference is in providing improved healthcare services to underserved populations. BCIs could be used to facilitate remote medical consultations and diagnostics, enabling individuals in remote areas to access healthcare professionals without the need to travel long distances. This has the potential to address the issue of limited healthcare infrastructure in developing countries, improving the overall quality and accessibility of healthcare services. BCIs could also play a pivotal role in enhancing educational opportunities in developing countries by providing innovative ways of learning and communication. BCIs could be used to assist individuals with disabilities in accessing educational resources and participating in online learning platforms. This can help bridge the digital divide and ensure that all individuals, regardless of their physical abilities, have equal access to education. BCIs could open up new possibilities for interactive and immersive learning experiences, ultimately enhancing the quality of education in developing countries. The application of BCIs in developing countries could also extend to improving communication and connectivity for marginalized communities. By enabling individuals with physical limitations to communicate more effectively through neural signals, BCIs could enhance social inclusion and empower individuals to actively participate in society. This has the potential to break down barriers and create a more inclusive and equitable society where everyone has the opportunity to engage and contribute to the community. The adoption of BCIs in developing countries could revolutionize healthcare, education, and communication, leading to a more interconnected and equal society.

## Neurotechnology for Mental Health Support

As neurotechnology continues to advance, the potential for using BCIs for mental health support is an area of growing interest and research. BCIs have the capability to monitor brain activity and provide real-time feedback, which could be instrumental in detecting and managing mental health conditions such as anxiety, depression, and PTSD. By utilizing neurotechnology, therapists and clinicians may gain valuable insights into the brain's functioning and tailor interventions more effectively based on individual neurological responses. This personalized approach could significantly enhance the efficacy of mental health treatments and lead to better outcomes for patients. One key advantage of incorporating neurotechnology into mental health support is the potential for early intervention and preventive measures. BCIs have the ability to detect subtle changes in brain activity that may indicate the onset of a mental health issue before symptoms manifest overtly. This early detection could pave the way for timely interventions, such as targeted therapies or neurofeedback training, to address underlying neural imbalances and prevent the escalation of mental health challenges. By leveraging neurotechnology for proactive monitoring and intervention, individuals may have a better chance of maintaining mental well-being and resilience against stressors or triggers that could exacerbate psychological distress. The integration of BCIs in mental health support holds promise for enhancing the accessibility and effectiveness of therapy for diverse populations. Through remote monitoring and telehealth platforms, individuals in remote or underserved areas can potentially access neurotechnological interventions and receive timely support for their mental health needs. BCIs could bridge communication gaps for individuals with conditions like autism or severe mental health disorders, enabling them to express their thoughts and emotions more effectively. By democratizing access to advanced mental health support through neurotechnology, the field of mental health care could witness a paradigm shift towards more personalized, effective, and inclusive interventions.

## Ethical Challenges in Implementing Neurotechnological Solutions

One of the most prominent ethical challenges in implementing neurotechnological solutions such as BCIs is the issue of privacy and security of neural data. As BCIs collect and transmit sensitive information directly from the brain, there is a significant risk of unauthorized access or misuse of this data. Protecting the confidentiality and integrity of neural data is crucial to ensure the trust and acceptance of BCI technology among users. The potential for data breaches or hacking presents a serious ethical dilemma that must be addressed through robust encryption protocols and secure data storage systems. Another ethical concern that arises in the implementation of neurotechnological solutions is the risk of technology dependence and abuse. As BCIs become more integrated into everyday life, there is a potential for individuals to become overly reliant on these devices for communication, entertainment, or even decision-making. Excessive dependence on BCI technology could lead to reduced autonomy and agency, raising questions about the ethical implications of such reliance on external devices for cognitive functions. It is essential to establish guidelines and frameworks to promote responsible and mindful use of BCIs while safeguarding against the misuse or overreliance on these technologies. The direct manipulation of the human brain through BCIs raises significant ethical considerations regarding autonomy, consent, and the potential for coercion. The ability to alter neural activity and influence cognitive processes through external interfaces raises complex questions about individual agency and the boundaries of cognitive freedom. Ethical frameworks must be established to delineate the limits of permissible interventions through BCIs and ensure that users retain control over their own thoughts and actions. Addressing these ethical challenges is essential to fostering a responsible and ethical approach to the implementation of neurotechnological solutions and supporting the positive impact of BCIs on society.

# XIX. NEUROTECHNOLOGY AND NEURODIVERSITY

Neurotechnology has the potential to revolutionize the way we understand and interact with the brain, particularly in the context of neurodiversity. By bridging the gap between technology and neuroscience, BCIs offer a unique platform for individuals with diverse neurological profiles to communicate and engage with the world in new ways. This has profound implications for the inclusion and empowerment of neurodiverse populations, as BCIs can provide alternative means of expression and connection that are tailored to individual needs and abilities. One of the key advantages of neurotechnology in the context of neurodiversity is its ability to support and enhance communication for individuals with diverse neurological profiles. For those who may face challenges in traditional modes of communication, such as verbal or written language, BCIs offer a direct pathway for expressing thoughts, emotions, and desires. This not only facilitates greater autonomy and agency for neurodiverse individuals but also promotes a deeper understanding and appreciation of their unique perspectives and experiences. The use of neurotechnology in the context of neurodiversity highlights the importance of designing inclusive and accessible technologies that cater to a wide range of cognitive and sensory abilities. By prioritizing user-centered design principles and embracing the diversity of human cognition, neurotechnologies like BCIs have the potential to break down barriers and create more equitable opportunities for all individuals, regardless of their neurological differences. This shift towards a more inclusive and neurodiverse-friendly technology landscape represents a significant step towards building a more accepting and supportive society for all individuals, regardless of their neurocognitive profiles.

## Enhancing Accessibility for Neurodiverse Individuals

As technology continues to advance, there is a growing need to enhance accessibility for neurodiverse individuals, particularly in the realm of neurotechnology and BCIs. By improving accessibility, individuals with conditions such as autism, ADHD, and dyslexia can better engage with technology and access its benefits. One way to enhance accessibility is by designing user interfaces that cater to the unique needs of neurodiverse individuals, such as providing customizable settings for sensory sensitivities or alternative modes of input for those with motor impairments. Incorporating universal design principles can also help ensure that technology is usable by a wide range of individuals, including those with neurodiverse abilities. Fostering a more inclusive design approach in the development of neurotechnology can lead to greater participation and engagement from neurodiverse individuals in various aspects of society, including education, employment, and social interactions. By creating neurotechnology devices that support different learning styles or communication preferences, individuals with neurodiverse characteristics can more easily navigate their environments and participate in activities that were previously challenging. This inclusivity can also help reduce stigma and promote acceptance of neurodiversity in society, leading to a more supportive and understanding community for all individuals. In addition to improving accessibility through design considerations, it is crucial to provide adequate training and support for neurodiverse individuals to effectively utilize neurotechnology and BCIs. By offering tailored training programs and resources, individuals can learn how to leverage these technologies to enhance their cognitive abilities, communication skills, and overall quality of life. Investing in research and development focused on neurodiverse populations can lead to innovative solutions and advancements that address the unique needs and preferences of these individuals. Enhancing accessibility for neurodiverse individuals in the realm of neurotechnology is not only a matter of social responsibility but also a means of promoting diversity, inclusion, and empowerment in the digital age.

## Supporting Neurodivergent Communities with BCI Technology

As the field of neurotechnology continues to advance, there is a growing interest in how BCIs can support neurodivergent communities. By harnessing the power of BCI technology, individuals with conditions such as autism, ADHD, dyslexia, and other neurodevelopmental disorders may be able to enhance their communication, cognitive abilities, and overall quality of life. This is particularly important as traditional methods of support may not always be effective for those with neurodivergent traits, highlighting the need for innovative solutions that cater to individual differences in neurological functioning. One key benefit of utilizing BCI technology to support neurodivergent communities is the ability to customize interventions and therapies based on the unique needs of each individual. By analyzing real-time brain activity, BCIs can provide personalized feedback and interventions that target specific cognitive functions or behaviors. This level of precision and individualization can lead to more effective and efficient support strategies that empower individuals to reach their full potential. The non-invasive nature of many BCI technologies makes them accessible to a wide range of individuals, reducing barriers to entry and increasing the reach of neurodiversity support services. The integration of BCI technology in supporting neurodivergent communities can foster a more inclusive and understanding society. By promoting awareness and acceptance of neurodiversity, BCIs can help bridge communication gaps and foster empathy among individuals with varying cognitive profiles. This not only benefits neurodivergent individuals but also contributes to a more compassionate and inclusive society that values diversity in neurological functioning. The potential of BCI technology to support neurodivergent communities highlights the transformative power of neurotechnology in enhancing human well-being and promoting neurodiversity acceptance.

## Ethical Considerations in Neurotechnological Inclusivity

One of the key ethical considerations in neurotechnological inclusivity is the issue of informed consent. As BCIs become more advanced and widespread, ensuring that individuals fully understand the risks and implications of using such technology is imperative. Informed consent involves not only providing information about the potential benefits and drawbacks of BCI use but also ensuring that individuals have the capacity to make autonomous decisions regarding their participation in such technologies. This becomes particularly challenging when considering populations with cognitive impairments or vulnerabilities who may not be able to fully grasp the complexities of BCI technology. Another ethical concern in neurotechnological inclusivity is the potential for discrimination and exclusion. As BCIs may enhance cognitive abilities or provide individuals with new ways to communicate or interact with the world, those who do not have access to or cannot afford such technologies may be left behind. This raises questions about equity and social justice in a society where neurotechnology could potentially exacerbate existing inequalities. It is crucial to consider how to ensure that BCIs are accessible to all individuals, regardless of socioeconomic status, education level, or other factors that could create disparities in access to neurotechnological advancements. The ethical implications of data privacy and security in the context of BCIs cannot be understated. With BCIs directly interfacing with the human brain, the potential for invasive data collection and surveillance raises serious concerns about individual autonomy and the protection of sensitive neural information. Safeguards must be put in place to protect the privacy of neural data, prevent unauthorized access or misuse of such information, and establish clear guidelines on how neural data can be used ethically and responsibly. Addressing these ethical challenges is essential in ensuring that neurotechnological inclusivity is achieved in a way that respects the rights and dignity of all individuals involved.

# XX. NEUROTECHNOLOGY AND NEUROETHICS IN EDUCATION

As the field of neurotechnology continues to advance, its applications in education are becoming increasingly relevant. By utilizing BCIs in educational settings, educators can gain valuable insights into students' cognitive processes and engagement levels. This data can then be used to tailor instructional strategies to individual learning styles, ultimately enhancing the effectiveness of teaching and learning. Neurotechnology in education can provide opportunities for more personalized and adaptive learning experiences, leading to improved academic outcomes for students. One key benefit of incorporating neurotechnology in education is the potential to foster a deeper understanding of neuroplasticity and brain development in learners. By tracking brain activity and responses during learning tasks, educators can identify patterns of cognitive growth and areas for improvement. This information can guide the implementation of targeted interventions and educational strategies that support students' cognitive development and academic success. Neurotechnology can help educators identify early signs of learning difficulties or cognitive impairments, enabling timely interventions and support for at-risk students. The integration of neurotechnology in education also raises important ethical considerations related to privacy, data security, and informed consent. Safeguarding students' neural data and ensuring that it is used responsibly and ethically is paramount in educational settings. Careful consideration must be given to the potential unintended consequences of relying too heavily on neurotechnology in education, such as reinforcing inequities or detracting from other important aspects of learning. By navigating these ethical challenges thoughtfully and responsibly, educators can harness the transformative potential of neurotechnology to create more inclusive, personalized, and effective learning environments for all students.

## Implementing BCI in Educational Settings

The integration of BCIs in educational settings holds great promise for enhancing the learning experience for students. By utilizing BCI technology, educators can gain valuable insights into students' cognitive processes, attention levels, and engagement with learning materials. This data can be used to tailor instruction to individual learning styles and preferences, ultimately leading to more personalized and effective learning outcomes. BCI can provide real-time feedback to both teachers and students, allowing for immediate adjustments and interventions to optimize learning experiences in the classroom. Implementing BCI in educational settings can open up new avenues for students with physical disabilities or limitations. By enabling these students to interact with technology and learning materials using their brain signals, BCI can offer a more inclusive and accessible learning environment. This inclusivity can foster a sense of empowerment and independence among students who may face traditional barriers to learning. BCI technology can enhance communication and collaboration among students, allowing for seamless integration of diverse perspectives and capabilities in the classroom setting. The integration of BCI in educational settings represents a significant step towards revolutionizing the way we learn and teach. From personalized instruction to inclusive learning environments, the potential benefits of incorporating BCI technology in education are vast. It is essential to address ethical considerations, such as data privacy and security, as well as ensure that BCI technology is used responsibly and ethically in educational settings. The successful implementation of BCI in education has the power to transform the future of learning and pave the way for a more inclusive and innovative educational landscape.

## Enhancing Learning and Cognitive Development with Neurotechnology

Neurotechnology has the potential to revolutionize learning and cognitive development by enhancing the way individuals interact with technology. Through the use of BCIs, researchers are exploring new ways to improve memory, attention, and overall cognitive function. By directly connecting the brain to external devices, neurotechnology can facilitate personalized learning experiences tailored to each individual's unique neural patterns and preferences. This personalized approach can lead to more efficient learning outcomes and increased engagement, ultimately enhancing cognitive development in ways previously thought impossible. The integration of neurotechnology in educational settings has shown promising results in enhancing learning outcomes for individuals with cognitive disabilities or learning difficulties. BCIs can provide real-time feedback and adaptive learning experiences, allowing students to learn at their own pace and in a way that suits their individual learning styles. Neurotechnology can aid in the development of neurofeedback techniques to improve attention, focus, and memory retention, offering new possibilities for enhancing cognitive development in both academic and professional settings. The use of neurotechnology and BCIs in enhancing learning and cognitive development represents a groundbreaking advancement with immense potential for transforming education and cognitive enhancement. As researchers continue to explore the capabilities of these technologies, it is crucial to address ethical considerations, privacy concerns, and accessibility issues to ensure that the benefits of neurotechnology are equitably distributed. By harnessing the power of neurotechnology in educational and cognitive development settings, we can unlock new avenues for personalized learning, cognitive enhancement, and expanded human potential in the digital age.

## Ethical Implications of Neuroenhancement in Education

Neuroenhancement in education through the use of neurotechnology and BCIs poses a myriad of ethical implications that must be carefully examined. One of the central concerns is the potential for exacerbating existing educational disparities, as those with access to neuroenhancement technologies may gain unfair advantages over their peers. This could lead to widening inequalities in academic performance, creating a two-tiered education system based on neuroenhancement capabilities. The pressure to excel academically through neuroenhancement could negatively impact students' mental health and well-being, fostering a competitive and stressful learning environment. The use of neuroenhancement in education raises questions about autonomy and consent, particularly when it comes to minors whose parents may make decisions about enhancing their cognitive abilities without their full understanding or consent. The issue of agency and personal identity also comes into play, as the use of neuroenhancement technologies could blur the lines between natural ability and artificially enhanced performance, potentially altering individuals' sense of self and self-worth. Concerns about data privacy and security arise, as neuroenhancement technologies involve the collection and storage of sensitive neural data, raising questions about who has access to this information and how it may be used for commercial or surveillance purposes. In addressing the ethical implications of neuroenhancement in education, policymakers, educators, and technologists must work together to establish clear guidelines and regulations to ensure equitable access, informed consent, and data protection. Ethical frameworks should be developed to guide the responsible implementation of neuroenhancement technologies in educational settings, emphasizing the importance of transparency, equity, and respect for individual autonomy. Public dialogue and engagement are crucial in fostering a deeper understanding of the ethical dilemmas posed by neuroenhancement in education, encouraging critical reflection and informed decision-making to navigate the complex terrain of enhancing cognitive abilities through neurotechnology.

# XXI. NEUROTECHNOLOGY AND EMOTIONAL INTELLIGENCE

In the realm of neurotechnology, the intersection between BCIs and emotional intelligence offers a fascinating avenue for exploration. Understanding how neurotechnology can enhance emotional intelligence has the potential to revolutionize the way humans interact with technology and each other. By tapping into the brain's intricate mechanisms for processing emotions, BCIs could enable more nuanced and empathetic communication in virtual environments, bridging the gap between the digital and physical worlds. This fusion of technology and emotion holds promise for applications in fields such as mental health therapy, virtual reality experiences, and even interpersonal communication. Recent advancements in neurotechnology have shown promising results in decoding emotional states based on brain activity. By harnessing the power of BCIs to detect and interpret emotional cues, individuals may be able to express their feelings in a more authentic and nuanced manner. This enhanced emotional communication could lead to more meaningful and empathetic interactions in both personal and professional settings. The ability to detect emotional responses in real-time through neurotechnology could revolutionize the way we understand and respond to the emotions of others, fostering deeper connections and reducing misunderstandings. As neurotechnology continues to evolve, researchers are increasingly exploring the potential of BCIs to enhance emotional intelligence through real-time feedback and personalized interventions. By leveraging the insights gained from decoding neural signals associated with emotions, individuals could be supported in developing greater self-awareness and empathy. This advancement in neurotechnology could pave the way for a future where emotional intelligence is not only enhanced but also democratized, creating a more empathetic and connected society. The fusion of neurotechnology and emotional intelligence has the potential to redefine how we relate to ourselves and others, ushering in a new era of empathetic and emotionally intelligent human-computer interactions.

## BCI Applications in Emotion Recognition

The field of neuroscience has made significant progress in recent years, leading to the development of innovative technologies such as BCIs. One of the exciting applications of BCIs is in the recognition of human emotions. By analyzing brain activity, BCIs can detect patterns associated with various emotions, providing a valuable tool for enhancing human-computer interaction. Emotion recognition using BCIs holds immense potential in fields such as healthcare, gaming, and virtual reality, where understanding user emotions can significantly improve the overall experience. In healthcare, BCI-based emotion recognition can revolutionize the way we diagnose and treat mental health disorders. By accurately detecting emotional states through brain signals, clinicians can tailor interventions to individual patients, leading to more personalized and effective treatment approaches. In the gaming industry, emotion recognition can create immersive experiences by adapting gameplay to the player's emotional state in real-time. This not only enhances user engagement but also opens up new possibilities for developing emotionally intelligent game systems that respond to the player's feelings. In virtual reality applications, emotion recognition using BCIs can enhance the sense of presence and realism by adjusting the virtual environment based on the user's emotional responses. This can lead to more engaging and impactful virtual experiences, whether used for training simulations, therapeutic purposes, or entertainment. The integration of emotion recognition with BCIs represents a promising avenue for unlocking the full potential of human-computer interaction, offering exciting opportunities for creating more empathetic and responsive technologies.

## Enhancing Emotional Regulation through Neurofeedback

Neurofeedback is a promising technique that aims to enhance emotional regulation by providing real-time feedback on brain activity. By training individuals to modulate their neural patterns, neurofeedback can help improve emotional resilience and reduce symptoms of anxiety, depression, and other mood disorders. Studies have shown that neurofeedback can lead to significant improvements in emotional regulation by promoting self-awareness and self-regulation skills. This form of brain training has also been linked to changes in brain connectivity and activity, suggesting long-lasting effects on emotional processing. Neurofeedback can be particularly beneficial for individuals with conditions such as ADHD, PTSD, and autism, where emotional dysregulation is a common symptom. By targeting specific brain regions associated with emotional control, neurofeedback can help rewire neural circuits and promote healthier emotional responses. This personalized approach to mental health treatment offers a non-invasive and potentially effective way to address underlying neurobiological factors contributing to emotional dysregulation. The ability of neurofeedback to target specific brain networks implicated in emotional processing allows for a more targeted and efficient intervention compared to traditional therapies. The field of neurofeedback holds significant promise in enhancing emotional regulation and overall mental well-being. As technology continues to advance, incorporating neurofeedback into therapeutic practices could revolutionize the way we approach emotional dysregulation and mental health disorders. By harnessing the power of neuroplasticity and BCIs, individuals can learn to regulate their emotions more effectively, leading to improved quality of life and psychological resilience. The integration of neurofeedback into mainstream mental health interventions could pave the way for a future where emotional regulation is not just a goal but a tangible reality for many individuals.

## Ethical Considerations in Emotional Manipulation with Neurotechnology

One of the key ethical considerations in emotional manipulation with neurotechnology is the potential for infringement upon an individual's autonomy and agency. By directly interfacing with the brain and influencing emotional responses, there is a risk of overriding an individual's natural thought processes and emotions, leading to questions of consent and control. This raises concerns about the ethical implications of manipulating emotions without the explicit consent of the individual, especially in cases where the technology is used for commercial or manipulative purposes. The issue of transparency and accountability is also crucial in the ethical use of neurotechnology for emotional manipulation. It is essential for developers and users of these technologies to be transparent about the capabilities and limitations of the devices, as well as the potential implications of emotional manipulation. Mechanisms for accountability and oversight need to be established to ensure that the technology is used responsibly and ethically, with safeguards in place to prevent misuse or abuse. Considerations of equity and social justice must be taken into account when exploring emotional manipulation with neurotechnology. There is a risk that such technology could disproportionately impact vulnerable populations or exacerbate existing social inequalities. It is crucial to consider the potential consequences of emotional manipulation on individuals from diverse backgrounds and to ensure that ethical guidelines are in place to promote fairness and equality in the application of this technology. Navigating the ethical considerations in emotional manipulation with neurotechnology requires a thoughtful and nuanced approach that prioritizes respect for individual autonomy, transparency, accountability, and social justice.

# XXII. NEUROTECHNOLOGY IN LAW ENFORCEMENT AND CRIMINAL JUSTICE

Neurotechnology has made significant advancements in recent years, particularly in the field of law enforcement and criminal justice. The integration of BCIs in these sectors has the potential to revolutionize how investigations are conducted and how suspects are interrogated. By leveraging neurotechnology, law enforcement agencies can enhance their ability to gather evidence, detect deception, and even predict criminal behavior. This technology can provide insights into the cognitive processes of individuals, allowing for more effective and efficient investigative procedures. One key application of neurotechnology in law enforcement is the use of brain imaging techniques such as fMRI to detect patterns of neural activity associated with deception. By analyzing brain activity during questioning, investigators can identify signs of deceit and use this information to guide their inquiries. This approach has the potential to improve the accuracy of lie detection and reduce the risk of false confessions. BCI technology can be utilized to decipher brain signals related to memory retrieval, potentially aiding in the identification of suspects and the reconstruction of crime scenes. Despite the promising applications of neurotechnology in law enforcement, ethical and legal concerns must be carefully considered. The use of BCI for interrogation purposes raises questions about privacy, consent, and the potential for coercive use of the technology. Safeguards must be put in place to ensure that neurotechnology is deployed ethically and in accordance with established legal standards. By addressing these challenges, society can harness the full potential of neurotechnology in law enforcement while upholding fundamental principles of justice and respect for individual rights.

## BCI for Lie Detection and Interrogation

As BCIs continue to advance, there is growing interest in utilizing this technology for lie detection and interrogation purposes. By tapping into patterns of brain activity, BCIs have the potential to provide a more accurate and reliable means of detecting deception compared to traditional methods such as polygraph tests. This could revolutionize the field of forensic psychology and criminal investigations, offering a more objective and scientific approach to uncovering the truth. One of the key advantages of using BCIs for lie detection is the direct access they provide to the brain's signals, bypassing the need for verbal or physical responses that can be manipulated by the subject. By analyzing neural activity associated with deception, researchers can develop algorithms that distinguish between truthful and deceptive responses with a high degree of accuracy. This could lead to more efficient and precise interrogation techniques, minimizing the risk of false confessions or wrongful convictions. The use of BCIs for lie detection also raises significant ethical and legal concerns. Issues relating to privacy, consent, and the reliability of the technology must be carefully considered before widespread implementation in legal settings. There is a risk of intrusive governmental surveillance if BCIs are used in interrogation processes without proper oversight and regulation. As researchers continue to explore the potential of BCIs for lie detection, it is crucial to address these ethical challenges to ensure the responsible and ethical use of this technology in the pursuit of justice.

## Neuroimaging in Criminal Profiling and Evidence Analysis

Neuroimaging techniques have revolutionized criminal profiling and evidence analysis by providing invaluable insights into the workings of the human brain. By utilizing technologies such as fMRI and EEG, investigators can uncover unique patterns of brain activity associated with specific cognitive processes. This has proven to be especially useful in determining deception, memory retrieval, and emotional responses in suspects and witnesses, aiding in the identification of crucial information for criminal investigations. FMRI has been used to detect deception by analyzing brain regions associated with decision-making and cognitive control, offering a new level of accuracy in detecting deceit. Neuroimaging has the potential to offer objective and scientifically grounded evidence in criminal cases, reducing the reliance on subjective assessments or witness testimonies. By examining brain activity patterns in response to stimuli or tasks relevant to a crime scene, forensic experts can gain insights into the mental processes of individuals involved. This can help corroborate or challenge witness statements, assess the credibility of testimonies, and provide additional evidence to support or refute allegations in court. The use of neuroimaging in criminal profiling adds a new dimension to evidence analysis, enhancing the validity and reliability of investigative procedures. Despite the promising advancements in the field, ethical considerations regarding the use of neuroimaging in criminal investigations must be carefully addressed. Issues such as privacy, consent, and the potential misuse of neuroimaging data raise important questions about the boundaries of using such technology in legal contexts. Concerns regarding the reliability and interpretability of neuroimaging results in court settings highlight the need for standardization and validation of these techniques within the criminal justice system. Neuroimaging holds great promise in enhancing criminal profiling and evidence analysis, but ongoing research and ethical considerations are essential to ensure its responsible and effective use in legal contexts.

## Ethical Challenges in the Use of Neurotechnology in Legal Contexts

When considering the ethical challenges in the use of neurotechnology in legal contexts, one central issue that arises is the potential for invasion of privacy. As neurotechnology advances, the ability to access and interpret neural data raises concerns about the protection of individuals' private thoughts and emotions. In legal settings, the use of neurotechnology for interrogation or lie detection purposes could infringe upon individuals' rights to keep their thoughts and intentions confidential. This poses a significant ethical dilemma regarding the boundaries of privacy and the extent to which neural information can be used in legal proceedings. The ethical implications of manipulating or altering neural processes through neurotechnology add another layer of complexity to its use in legal contexts. The idea of directly influencing brain activity raises questions about autonomy and free will. If individuals can be coerced or manipulated through neurotechnology, it challenges traditional notions of responsibility and accountability in legal systems. The potential for misuse or abuse of neurotechnology to control behavior or manipulate decisions further highlights the importance of considering the ethical implications of its applications in legal settings. The issue of equity and justice when utilizing neurotechnology in legal contexts is a critical ethical concern. Access to advanced neurotechnology may not be equally available to all individuals, potentially creating disparities in how neural data is collected and used in legal proceedings. This raises questions about fairness and impartiality in the legal system, as well as concerns about the potential for discrimination based on neural information. Addressing these ethical challenges requires careful consideration of the implications of using neurotechnology in legal contexts and the development of clear guidelines to ensure the responsible and ethical use of this powerful technology.

# XXIII. NEUROTECHNOLOGY AND WORKPLACE PRODUCTIVITY

Neurotechnology has the potential to significantly impact workplace productivity by enhancing cognitive abilities and optimizing human-machine interactions. One key aspect of this technology is the development of BCIs, which allows for direct communication between the brain and external devices. By utilizing BCIs in the workplace, individuals can seamlessly control computer applications, devices, and machinery through their thoughts alone, streamlining processes and improving overall efficiency. This level of interface can revolutionize tasks that require high levels of concentration and attention to detail, ultimately leading to increased productivity and output in various industries. The integration of neurotechnology in the workplace can also enhance employee well-being and mental health. By utilizing BCIs for tasks that require mental focus, individuals can avoid cognitive overload and reduce the risk of burnout. BCI technology can provide real-time feedback on an individual's cognitive state, allowing for personalized interventions to improve focus, stress management, and overall mental wellness. This aspect of neurotechnology can create a conducive work environment that prioritizes employee health and productivity, ultimately leading to a more satisfied and engaged workforce. The implementation of neurotechnology in the workplace also raises ethical concerns regarding privacy, data security, and potential misuse of cognitive data. Employers must establish strict protocols for data collection and storage to protect the sensitive information accessed through BCIs. Safeguards must be put in place to ensure that employees retain control over their cognitive data and that consent is obtained for any data collection or analysis. By addressing these ethical challenges, workplaces can leverage the benefits of neurotechnology while prioritizing the well-being and rights of their employees, ultimately creating a more productive, ethical, and sustainable work environment.

## BCI for Enhancing Focus and Task Performance

One significant application of BCIs is their potential to enhance focus and task performance. By utilizing EEG signals to monitor brain activity, BCIs can provide real-time feedback to help individuals maintain concentration and improve cognitive performance. This technology holds great promise in various fields, including education, professional development, and even sports training. Students could use BCIs to improve their focus during study sessions, professionals could enhance their productivity at work, and athletes could optimize their performance during training sessions. BCIs work by detecting and interpreting brain signals to control external devices or provide feedback to users. In the context of enhancing focus and task performance, BCIs can monitor brain waves associated with attention and concentration, providing users with immediate feedback on their mental state. By leveraging this information, BCIs can help individuals maintain focus, avoid distractions, and stay engaged in their tasks for more extended periods. BCIs can be tailored to specific goals or objectives, enabling users to customize their experience and optimize their performance in various activities. The integration of BCIs for enhancing focus and task performance represents a significant step towards unlocking the full potential of the human brain. By leveraging advanced technology to monitor and modulate cognitive processes, BCIs offer a powerful tool for improving productivity, learning efficiency, and overall mental performance. As research and development in this field continue to advance, we can expect to see further innovations that harness the power of BCIs to help individuals achieve their goals and excel in various domains. The incorporation of BCIs in daily life has the potential to revolutionize how we interact with technology and enhance our cognitive abilities for a brighter future.

## Neurofeedback Training for Stress Management in the Workplace

One of the applications of neurotechnology that has garnered significant interest in recent years is the use of neurofeedback training for stress management in the workplace. This approach involves using real-time brain activity monitoring to provide individuals with feedback on their stress levels and teach them techniques to better regulate their responses. By enabling employees to gain awareness of their physiological reactions to stress, neurofeedback training can help them develop coping strategies and improve their overall well-being in high-pressure work environments. This personalized approach to stress management has the potential to enhance productivity and reduce employee burnout, ultimately benefiting both individuals and organizations. Research studies have shown promising results for the effectiveness of neurofeedback training in reducing stress and improving mental health outcomes. By targeting specific brain regions associated with stress responses, such as the amygdala and prefrontal cortex, individuals can learn to self-regulate their emotional reactions and achieve a state of balance and relaxation. The use of neurofeedback in the workplace can empower employees to take control of their stress levels and cultivate resilience in the face of challenges. This proactive approach to mental health management aligns with the growing emphasis on holistic well-being in corporate settings, highlighting the importance of addressing the root causes of stress through innovative interventions like neurofeedback training. Incorporating neurofeedback training into workplace wellness programs represents a forward-thinking strategy for promoting employee health and performance. By harnessing the power of neurotechnology to improve stress management skills, organizations can create a supportive and inclusive work environment that prioritizes mental well-being. As the field of neurotechnology continues to advance, further research and implementation of neurofeedback training in the workplace could lead to a paradigm shift in how we approach stress management and mental health support in professional settings. Embracing such cutting-edge solutions reflects a commitment to fostering a positive and sustainable work culture that benefits both individuals and the overall success of the organization.

## Ethical Considerations in Employee Monitoring with Neurotechnology

When considering the ethical implications of employee monitoring with neurotechnology, one crucial aspect to evaluate is the concept of privacy. Neurotechnology allows for the direct monitoring and analysis of an individual's brain activity, raising concerns about the invasion of privacy and the potential for misuse of sensitive neural data. Employers must address the ethical dilemma of balancing the need for monitoring productivity and performance with respecting the privacy rights and autonomy of their employees. Striking a balance between these competing interests is essential to ensuring ethical practices in the workplace. Another significant ethical consideration in employee monitoring with neurotechnology is the issue of informed consent. Informed consent requires individuals to fully understand the purpose, risks, and implications of the monitoring process before agreeing to participate. With the complex nature of neurotechnology and the potential impact on individuals' cognitive and emotional privacy, obtaining informed consent becomes a crucial ethical requirement. Employees must have the right to make an informed decision about whether they are comfortable with their neural data being monitored and analyzed for work-related purposes. The potential for discrimination and bias in employee monitoring with neurotechnology poses an ethical challenge that requires careful consideration. Neural data may reveal sensitive information about an individual's cognitive processes, emotions, or mental health status, which could be used to make decisions regarding promotions, job assignments, or retention. Employers must establish clear guidelines and ethical frameworks to prevent discrimination based on neural data and ensure that monitoring practices do not lead to unfair or biased treatment of employees. Addressing these ethical considerations is vital in fostering a culture of trust, transparency, and respect in the workplace when implementing neurotechnology for employee monitoring.

# XXIV. NEUROTECHNOLOGY AND PERSONALIZED MEDICINE

Neurotechnology has the potential to revolutionize personalized medicine by offering tailored healthcare solutions based on individual brain activities and responses. Through the use of BCIs, healthcare providers can gather real-time data on a patient's neural patterns to create customized treatment plans. By analyzing neural signals, doctors can better understand a patient's condition and adjust therapies accordingly, leading to more effective and targeted interventions. This personalized approach can optimize treatment outcomes, minimize side effects, and improve overall patient wellbeing. One of the key advantages of incorporating neurotechnology into personalized medicine is the ability to monitor patients' progress and make real-time adjustments to their treatment plans. With BCI technology, healthcare providers can continuously assess a patient's neural activity to track changes in their condition and response to treatment. This dynamic approach allows for timely interventions and modifications, ensuring that patients receive the most effective and personalized care possible. By harnessing the power of neurotechnology, personalized medicine can become more adaptive, responsive, and tailored to the unique needs of each individual. The integration of neurotechnology in personalized medicine can enhance patient engagement and empowerment by providing individuals with greater insight into their own health and wellbeing. Through BCI technologies, patients can actively participate in their treatment process, monitor their progress, and make informed decisions about their care. This collaborative approach can lead to better patient outcomes, increased treatment adherence, and improved overall quality of life. By leveraging the capabilities of neurotechnology, personalized medicine can empower individuals to take control of their health and well-being, bridging the gap between patients and healthcare providers for a more holistic and patient-centered approach to healthcare.

## Precision Medicine Approaches with BCI Data

One promising application of BCIs data is in the field of precision medicine. By harnessing the power of neural signals and brain activity patterns, researchers can tailor medical treatments to individuals with unprecedented accuracy. BCI data can be used to create personalized therapy plans for patients with neurological disorders, such as epilepsy or Parkinson's disease. By analyzing the unique neural signatures of each patient, healthcare providers can optimize treatment outcomes and minimize side effects, leading to more effective and efficient care. Precision medicine approaches with BCI data can revolutionize the field of mental health by providing targeted interventions for individuals with psychiatric conditions. Researchers are exploring the use of BCI technology to develop personalized treatment strategies for depression, anxiety, and PTSD. By using real-time brain activity data to monitor patients' response to therapy, clinicians can adjust treatment protocols in real-time, optimizing outcomes and improving patient well-being. This individualized approach to mental health care has the potential to transform traditional treatment methods and improve the overall quality of life for individuals struggling with mental health challenges. The integration of BCI data into precision medicine approaches holds great promise for the future of healthcare. Imagine a world where medical treatments are tailored to each person's unique neural profile, ensuring maximum efficacy and minimal side effects. As technology continues to advance, the possibilities for using BCI data to revolutionize healthcare are endless. By leveraging the power of neural signals and brain activity patterns, researchers and healthcare providers can unlock new insights into human health and disease, leading to more personalized and effective treatment strategies. The potential of precision medicine approaches with BCI data is truly groundbreaking and has the potential to redefine the future of healthcare.

## Tailoring Treatment Plans Using Neurotechnological Insights

As technology continues to advance, the field of neurotechnology has emerged as a promising area for tailoring treatment plans using neurotechnological insights. By utilizing BCIs, researchers and healthcare professionals can gather valuable data on brain activity to customize treatment options for individuals with neurological disorders. This personalized approach is crucial in optimizing therapeutic outcomes and improving the quality of life for patients. With the ability to monitor and analyze neural signals in real-time, BCIs offer a novel way to tailor treatment plans that are specifically tailored to each individual's unique brain activity patterns. One of the key benefits of leveraging neurotechnological insights in treatment planning is the ability to target interventions based on precise neurological data. Through the use of BCIs, healthcare providers can gain a deeper understanding of how an individual's brain responds to various stimuli, allowing for more effective treatment strategies.

This personalized approach can lead to improved patient outcomes, reduced side effects, and enhanced treatment adherence. By incorporating neurotechnological insights into treatment planning, healthcare providers can enhance the precision and efficacy of interventions, ultimately improving the overall quality of care for individuals with neurological conditions. Tailoring treatment plans using neurotechnological insights can help bridge the gap between traditional healthcare practices and cutting-edge technology. By integrating BCIs into treatment protocols, healthcare providers can tap into the potential of neurotechnology to revolutionize the way neurological disorders are managed. This innovative approach not only facilitates more accurate diagnostics and treatment monitoring but also opens up new avenues for exploring novel therapeutic interventions. As research in neurotechnology continues to advance, the potential for tailoring treatment plans based on neurotechnological insights will only grow, offering new hope for individuals living with neurological conditions.

## Ethical Implications of Personalized Healthcare through Neurotechnology

One of the key ethical implications of personalized healthcare through neurotechnology is the issue of privacy and security concerning neural data. As BCIs collect and analyze sensitive information directly from the brain, there is a risk of unauthorized access and potential misuse of this data. Safeguarding the privacy of individuals and ensuring the security of neural data is crucial to maintaining ethical standards in the field of personalized healthcare. The possibility of neural data being used for commercial purposes without consent raises concerns about exploitation and manipulation of individuals' cognitive information. Another ethical challenge in the realm of personalized healthcare through neurotechnology is the potential risks of technology dependence and abuse. As BCIs become more integrated into medical treatments and everyday life, there is a risk of individuals becoming overly reliant on these technologies. This dependence can lead to a loss of autonomy and agency, raising questions about the ethical implications of relying on neurotechnology for basic functions and decision-making processes. The misuse or abuse of BCIs for unauthorized control or manipulation of individuals' neural activity raises ethical concerns about the boundaries of technological interventions in cognitive processes. The ethical implications of direct manipulation of the human brain through personalized healthcare using neurotechnology are significant. The ability to modulate neural activity and alter cognitive functions raises ethical questions about the potential impacts on personal identity, autonomy, and agency. The ethical consideration of altering individuals' brain functions through neurotechnology raises concerns about the boundaries of technological interventions in shaping human cognition. Addressing these ethical challenges is essential in ensuring that the development and application of personalized healthcare through neurotechnology are guided by ethical principles that prioritize the well-being and autonomy of individuals.

# XXV. NEUROTECHNOLOGY AND SOCIAL IMPACT

The field of neurotechnology has witnessed significant advancements in recent years, particularly in the development of BCIs. These interfaces have the potential to revolutionize the way humans interact with technology, enabling direct communication between the brain and external devices. As BCI technology continues to evolve, it holds the promise of enhancing human capabilities and opening up new possibilities for individuals with motor disabilities or neurological disorders. The integration of BCIs into everyday life could lead to a future where humans can control prosthetic limbs, communicate through thoughts alone, and even experience virtual reality through mind-controlled interfaces. One of the key aspects in the operation of BCIs lies in the intricate design of its components, including sensors, signal processing techniques, and decoding algorithms. These components work in tandem to translate neural signals into actionable commands for external devices, bridging the gap between the human brain and technology. The distinction between invasive and non-invasive BCIs plays a crucial role in determining the level of precision and invasiveness of the interface. While invasive BCIs offer higher resolution and control, non-invasive options provide a more accessible and user-friendly approach to brain-computer communication. Despite the promising applications of BCIs in fields such as medicine, entertainment, and communication, their widespread adoption raises ethical and social challenges that cannot be overlooked. Issues surrounding privacy, security, potential misuse of neural data, and the ethical implications of directly manipulating the human brain highlight the need for careful consideration and regulation of BCI technologies. Addressing these challenges is paramount to ensuring that the integration of neurotechnology into society is done responsibly and ethically, ultimately shaping a future where BCIs can coexist harmoniously with humanity, enhancing our quality of life without infringing on our rights and autonomy.

## Addressing Social Inequality through Access to Neurotechnological Advancements

Access to neurotechnological advancements has the potential to address social inequality by providing individuals with disabilities the opportunity to participate more fully in society. Individuals with physical disabilities can benefit from BCIs that allow them to control prosthetic limbs or assistive devices with their thoughts. By enabling individuals to interact with their environment in novel ways, neurotechnology can reduce barriers to employment, education, and social participation, ultimately promoting a more inclusive society. Improving access to neurotechnological advancements can also help bridge the digital divide that perpetuates social inequality. In a world where technology plays an increasingly central role in daily life, individuals who lack access to the latest advancements may find themselves at a disadvantage in terms of economic opportunities, education, and social connections. By ensuring that neurotechnological advancements are accessible to all, regardless of socio-economic status, we can level the playing field and create a more equitable society where everyone has the opportunity to thrive. Addressing social inequality through access to neurotechnological advancements requires a multi-faceted approach that considers not only the technological aspects but also the social, economic, and ethical implications of these advancements. By integrating principles of equity and inclusion into the development and deployment of neurotechnology, we can harness its transformative potential to create a more just and equal society. Through collaboration between researchers, policymakers, and communities, we can work towards a future where neurotechnological advancements help to break down barriers and create a more inclusive and equitable world for all.

## Promoting Empathy and Understanding with BCI Technology

In the realm of neuroscience and technology, the development of BCIs holds tremendous potential for promoting empathy and understanding among individuals. By allowing direct communication between the brain and external devices, BCIs have the capacity to bridge gaps in communication that may arise due to physical limitations or neurodiversity. Individuals with severe physical disabilities could use BCIs to express their thoughts and emotions, fostering a deeper understanding of their perspectives and enhancing their overall quality of life. The use of BCIs in immersive experiences, such as mind-controlled video games and virtual reality simulations, can provide neurotypical individuals with a unique opportunity to step into the shoes of others and gain a deeper sense of empathy. Through these experiences, users can develop a greater understanding of different cognitive processes and perspectives, leading to increased compassion and empathy towards individuals with diverse neurocognitive profiles. This immersive approach to learning and experiencing the world through another's eyes has the potential to reshape societal attitudes towards diversity and inclusion. The integration of BCI technology in various aspects of everyday life has the potential to revolutionize the way in which we interact with each other and understand the complexities of the human mind. By fostering empathy and promoting understanding through direct neural communication, BCIs can empower individuals to connect on a deeper level, transcending physical and cognitive barriers. As the field of neurotechnology continues to advance, it is essential to harness the transformative power of BCIs to create a more inclusive and empathetic society.

## Ethical Considerations in Shaping Social Norms through Neurotechnology

One ethical consideration in shaping social norms through neurotechnology is the potential for manipulation and control. As BCIs become more advanced, there is a risk that individuals or institutions could use this technology to influence or even dictate people's thoughts and actions. This raises concerns about issues such as autonomy, consent, and privacy. If social norms are shaped through neurotechnology without adequate safeguards in place, it could lead to a loss of individual agency and a lack of genuine freedom in decision-making processes. Another ethical consideration is the impact of neurotechnology on inequality and social justice. As with any technological advancement, there is a risk that BCIs could exacerbate existing disparities in society. If only certain groups have access to this technology, it could widen the gap between the haves and have-nots. There may be concerns about discrimination based on neural data, leading to further marginalization of already vulnerable populations. It is essential to consider how the development and implementation of neurotechnology could either promote or hinder social equity and fairness. Ethical considerations also extend to the potential cultural and societal implications of shaping social norms through neurotechnology. Different communities and cultures may have varying views on the appropriate uses of such technology, raising questions about cultural sensitivity and respect. It is crucial to engage in meaningful dialogue and collaboration with diverse stakeholders to ensure that the development and deployment of BCIs are done in a way that respects and upholds the values and beliefs of different social groups. Navigating the ethical considerations in shaping social norms through neurotechnology requires a careful balance between progress and responsibility to safeguard individual rights and societal well-being.

# XXVI. NEUROTECHNOLOGY AND NEUROPHILOSOPHY

Neurophilosophy, a branch of philosophy that explores the relationship between neuroscience and philosophical questions, has gained significant relevance with the advancements in neurotechnology, particularly BCIs. This interdisciplinary field delves into the ethical and philosophical implications of using technology to interface with the human brain, raising profound questions about identity, consciousness, and free will. As neurotechnology continues to progress, the intersection of neuroscience and philosophy becomes increasingly significant in understanding the impact of BCI on individuals and society. One key aspect of neurophilosophy in the context of BCI is the issue of agency and autonomy. The ability of individuals to control external devices using their thoughts challenges traditional notions of agency and free will. Philosophical debates on the nature of consciousness and the self are further amplified by the technological advancements that allow for direct communication between the brain and external systems. This intersection of technology and philosophy opens up new avenues for exploring the nature of the mind and the boundaries of human cognition. Neurophilosophy also grapples with questions of identity and the potential implications of merging human intelligence with artificial systems. The concept of neuroenhancement through BCI raises ethical concerns about altering human capabilities and the risks of creating unequal access to cognitive enhancements. As neurotechnology continues to evolve, the field of neurophilosophy serves as a critical lens through which to examine the societal, ethical, and existential implications of BCIs. By engaging in thoughtful discourse at the intersection of neuroscience and philosophy, we can navigate the complexities of advancing technology while upholding ethical principles and preserving the essence of humanity.

## Exploring Consciousness and Identity with BCI

The exploration of consciousness and identity through BCIs opens up a fascinating and complex realm of possibilities for understanding the human mind. By directly interfacing with the brain's electrical signals, BCIs provide a unique window into how thoughts are formed and translated into action. This technology allows researchers to delve into the nuances of consciousness, shedding light on what makes us who we are at a fundamental level. Through BCI, it becomes possible to unlock the mysteries of the mind and gain insights into the intricacies of human identity. One of the most intriguing aspects of utilizing BCI for exploring consciousness and identity is its potential to reveal the interconnectedness of cognitive processes. By analyzing patterns of brain activity, researchers can identify neural correlates of specific mental states and behaviors, offering a deeper understanding of how different aspects of consciousness are linked. This holistic approach to studying the mind through BCI paves the way for a more comprehensive perspective on the intricate web of thoughts, emotions, and memories that shape our sense of self. By uncovering the underlying mechanisms of consciousness, BCI has the power to revolutionize our understanding of identity and the human experience. The insights gained from exploring consciousness and identity with BCI have profound implications for fields beyond neuroscience, such as psychology, philosophy, and artificial intelligence. By deciphering the neural substrates of consciousness and identity, researchers can inform theories of self-awareness, decision-making processes, and the nature of subjective experience. This interdisciplinary approach not only enriches our understanding of the human mind but also lays the groundwork for developing more advanced AI systems that mimic human cognition. The exploration of consciousness and identity with BCI holds promises of transformative discoveries that could redefine our understanding of what it means to be human.

## Philosophical Implications of Direct Brain Manipulation

Ethical considerations arise when contemplating the direct manipulation of the human brain through neurotechnology. The ability to alter neural activity or control cognitive functions through BCIs raises profound philosophical questions about autonomy, identity, and the nature of self. If individuals can modify their thoughts or emotions with a simple technological intervention, what does this mean for our understanding of free will and personal agency? These questions challenge traditional notions of human consciousness and raise concerns about the ethical boundaries of neuroenhancement. The implications of direct brain manipulation extend beyond individual autonomy to societal impact. The potential for widespread adoption of BCIs raises questions about equity and access to cognitive enhancements. Will neurotechnology exacerbate existing societal disparities, creating a new divide between those who can afford cognitive enhancements and those who cannot? The use of BCIs for purposes such as memory manipulation or emotional regulation raises concerns about coercion and manipulation. The power to control one's thoughts or feelings through external devices could be exploited for nefarious purposes, leading to ethical dilemmas in areas such as informed consent and personal autonomy. The philosophical implications of direct brain manipulation through neurotechnology are multifaceted and complex. As we navigate the ethical challenges and societal impacts of BCIs, it is crucial to engage in thoughtful reflection and dialogue to ensure that these technologies are developed and utilized in a responsible and ethical manner. By addressing these philosophical questions and ethical dilemmas, we can strive towards a future where neurotechnology enhances the human experience while upholding fundamental principles of autonomy, privacy, and social justice.

## Ethical Considerations in Neurophilosophical Inquiry

In the realm of neurophilosophical inquiry, ethical considerations play a crucial role in guiding the development and utilization of BCIs. One key ethical concern revolves around the issue of privacy and security of neural data obtained through BCIs. As these devices have the capability to access and interpret an individual's brain signals, there is a risk of unauthorized access to sensitive information, leading to potential breaches of privacy. The storage and sharing of neural data raise concerns about data ownership and the possibility of misuse by third parties, highlighting the need for robust data protection measures and ethical guidelines. The ethical implications of direct manipulation of the human brain through BCIs raise questions about autonomy and agency. The ability to control external devices or even modify cognitive functions through neural interfaces blurs the line between individual identity and external influence. This brings into question the concept of personal autonomy, as individuals may be susceptible to external manipulation or coercion through the use of BCIs. Ethical frameworks must be developed to ensure that the use of these technologies respects the rights and freedoms of individuals, while also promoting the responsible and beneficial application of neurotechnology. The potential risks of technology dependence and abuse highlight the importance of promoting ethical practices in the development and deployment of BCIs. As these devices become more integrated into everyday life, there is a risk of over-reliance on technology for cognitive or physical tasks, which could lead to societal implications such as reduced human agency or loss of critical thinking skills. By addressing these ethical considerations in neurophilosophical inquiry, we can ensure that the advancement of BCI technology aligns with ethical standards and contributes positively to the well-being and autonomy of individuals within society.

# XXVII. NEUROTECHNOLOGY IN SPACE EXPLORATION

Neurotechnology has the potential to revolutionize space exploration by enhancing the connection between humans and machines in the harsh environment of outer space. BCIs can enable astronauts to control equipment, interact with spacecraft systems, and even communicate with Earth using just their thoughts. By utilizing BCI technology, space agencies can streamline operations, increase efficiency, and improve the overall safety of space missions. This could lead to significant advancements in our ability to explore and colonize other planets, as well as make long-duration space travel more manageable for astronauts. Neurotechnology can also play a crucial role in monitoring the physical and mental well-being of astronauts during extended space missions. BCIs can provide real-time feedback on the cognitive and emotional states of crew members, helping to identify signs of stress, fatigue, or other issues that could impact their performance. This data can be used to optimize mission planning, personalize support strategies, and ensure the overall health and resilience of the crew. By integrating neurotechnology into space exploration, we can enhance the capabilities of human spaceflight and pave the way for sustainable long-term missions beyond Earth's orbit. The development of neurotechnology for space exploration can have wider implications for terrestrial applications, such as healthcare, gaming, and communication. The advancements made in BCI technology to support astronauts in space can be adapted and refined for use in various industries on Earth, leading to innovative solutions for improving quality of life and enhancing human capabilities.

By investing in research and development in the field of neurotechnology, we not only propel space exploration forward but also unlock the potential for transformative technologies that benefit society as a whole.

## BCI for Astronaut Training and Monitoring

Advancements in BCI technology have paved the way for its potential application in astronaut training and monitoring. By utilizing BCI systems, astronauts can undergo mental training and cognitive assessments in a more streamlined and efficient manner. These systems can track brain activity during various tasks and simulations, providing valuable insights into an astronaut's cognitive performance and mental state. BCIs can be used to monitor stress levels and fatigue, crucial factors to consider in the extreme conditions of space travel. BCI technology offers the possibility of enhancing communication between astronauts and mission control. With real-time monitoring of brain activity, BCIs can enable instant transmission of crucial information and alerts, improving response times and decision-making processes. This level of connectivity can be crucial in emergency situations or when faced with unexpected challenges during space missions. BCIs can facilitate hands-free operation of equipment and interfaces, allowing astronauts to efficiently control equipment and systems with minimal physical effort. The integration of BCI technology in astronaut training and monitoring could revolutionize the way we prepare and support individuals in space exploration. By providing insights into mental states and cognitive performance, BCIs can help optimize training protocols and enhance overall mission success. The real-time monitoring capabilities of BCIs can improve safety measures and enhance overall communication strategies in space missions. As technology continues to advance, the potential for BCIs to play a significant role in astronaut training and monitoring is promising, offering a glimpse into a future where human-machine interaction reaches new heights in outer space.

## Enhancing Cognitive Performance in Extreme Environments

Research has shown that cognitive performance can be significantly impacted in extreme environments, such as high altitude, underwater, or space. In these challenging conditions, individuals may experience cognitive deficits, memory issues, and impaired decision-making abilities due to factors like limited oxygen, sensory deprivation, or high levels of stress. Enhancing cognitive performance in such environments has become a critical area of focus, with the goal of improving safety, efficiency, and overall mission success. Strategies to enhance cognitive performance in extreme environments include the use of neurotechnology, such as BCIs, to provide real-time feedback and cognitive augmentation. One promising approach to enhancing cognitive performance in extreme environments is through the use of BCIs to monitor and optimize brain activity. BCIs can detect changes in brain waves associated with cognitive functions like attention, memory, and decision-making, allowing individuals to receive immediate feedback on their cognitive state. By leveraging this technology, individuals operating in extreme environments can make quicker and more accurate decisions, leading to improved performance outcomes. BCIs can be used to provide cognitive support, such as alerting individuals to potential errors or guiding them through complex tasks, further enhancing cognitive performance in challenging conditions. BCIs can be tailored to individual users, taking into account their unique cognitive strengths and weaknesses to optimize performance in extreme environments. By utilizing personalized neurofeedback and cognitive training programs, BCIs can help individuals improve their cognitive abilities over time, resulting in enhanced performance under extreme conditions. This individualized approach to cognitive enhancement not only benefits individuals in extreme environments but also has broader implications for cognitive training and performance optimization in various professional and personal settings. The integration of BCIs in enhancing cognitive performance in extreme environments represents a promising avenue for improving human capabilities and achieving optimal performance in challenging conditions.

## Ethical Considerations in Space Travel with Neurotechnological Support

When considering the advancements in space travel with neurotechnological support, ethical considerations play a crucial role in ensuring the well-being of astronauts and the preservation of their autonomy. One key aspect to consider is the potential invasion of privacy with the use of BCIs in space missions. BCIs can capture and transmit neural data, raising concerns about the confidentiality and security of astronauts' thoughts and emotions. It becomes essential to establish strict protocols and regulations to safeguard the privacy of individuals in such settings, especially considering the isolated and confined nature of space exploration. The issue of consent becomes paramount when integrating neurotechnological support in space travel. Astronauts must fully understand and consent to the use of BCIs for monitoring their cognitive and emotional states, as well as for enhancing their performance in tasks. Informed consent is crucial to ensure that individuals are aware of the potential risks and benefits associated with neurotechnological interventions, allowing them to make autonomous decisions regarding their participation in space missions. Ethical guidelines must be established to guarantee that astronauts have the right to withdraw from any neurotechnological procedures if they wish, without facing repercussions or consequences. The potential for technological dependence and manipulation in space travel with neurotechnological support raises ethical concerns about the autonomy and agency of astronauts. As BCIs become more integrated into everyday activities, there is a risk of individuals becoming reliant on these technologies for decision-making and control. This dependence could erode the freedom of astronauts to make independent choices, potentially leading to ethical dilemmas regarding who has the power and authority in decision-making processes. It is imperative to address these ethical challenges by promoting a balance between the benefits of neurotechnological support and the preservation of astronauts' autonomy and agency in space exploration.

# XXVIII. NEUROTECHNOLOGY AND ARTISTIC INNOVATION

Neurotechnology presents a new frontier for artistic innovation, providing creators with tools to explore the depths of human cognition and expression. By interfacing with the brain, artists can tap into a realm of creativity that transcends traditional mediums, offering a unique avenue for immersive experiences and interactive artworks. The ability to translate neural activity into visual and auditory stimuli opens up a world of possibilities for engaging audiences on a profound level, blurring the lines between art and technology in ways never seen before. One of the most exciting aspects of neurotechnology in the realm of artistic innovation is its potential to democratize creativity, allowing individuals of all abilities to express themselves in new and meaningful ways. By bypassing traditional modes of communication and tapping directly into the neural substrate of the mind, neurotechnology offers a level playing field for artists with physical limitations or communication disorders. This inclusivity not only expands the reach of artistic expression but also challenges societal norms around ability and access, paving the way for a more diverse and equitable creative landscape. As neurotechnology continues to evolve and integrate into the artistic sphere, it raises important questions about the nature of creativity, authorship, and the boundaries of human expression. The collaboration between human minds and machine interfaces blurs the distinction between artist and instrument, challenging conventional notions of artistic agency and autonomy. In this dynamic interplay between technology and creativity, new forms of art emerge that push the boundaries of human perception and understanding, inviting audiences to reconsider their relationship with both art and technology in a rapidly evolving digital age.

## Brainwave Art and Neurofeedback in Creative Processes

Neurofeedback, often used in conjunction with brainwave art, has shown promise in enhancing creative processes. By providing real-time feedback on brain activity, individuals can learn to control their mental state and achieve a state of flow conducive to creativity. This technology allows users to see how their brainwaves change in response to various stimuli, enabling them to better understand their own cognitive processes and optimize them for creative endeavors. Through neurofeedback, individuals can train their brains to enter specific states associated with heightened creativity, such as increased alpha activity in the prefrontal cortex. The integration of brainwave art and neurofeedback in creative processes also offers a unique avenue for self-expression and exploration. By visualizing their brain activity in the form of art, individuals can gain insights into the abstract workings of their minds and uncover new sources of inspiration. This innovative approach not only enhances the creative experience but also allows for a deeper connection with one's cognitive processes. The use of brainwave art as a feedback mechanism can help individuals track their progress and monitor improvements in their creative abilities over time. The combination of brainwave art and neurofeedback presents an exciting opportunity to push the boundaries of creative expression and cognitive enhancement. By leveraging cutting-edge technology to tap into the intricacies of the brain, individuals can unlock new creative potential and foster a deeper understanding of their cognitive capabilities. As neurotechnology continues to advance, the integration of brainwave art and neurofeedback in creative processes has the potential to revolutionize how we approach art, innovation, and self-discovery in the digital age. This fusion of art and science holds promise for shaping the future of creativity and human expression.

## Enhancing Artistic Expression through BCI Technology

One of the most exciting developments in the field of neurotechnology is the potential for enhancing artistic expression through BCI technology. By allowing individuals to directly interface their thoughts and intentions with digital tools, BCIs offer a new frontier for creative expression. Artists can now explore innovative ways to create music, visual art, and even literature using only their minds, expanding the possibilities of what can be achieved in the realm of art. This direct neural interface provides a unique avenue for artists to push the boundaries of traditional forms and develop new modes of expression that were previously unimaginable. Through the use of BCI technology, artists can tap into the raw essence of their creative impulses and translate them into tangible works of art without the constraints of physical tools or mediums. This opens up a world of possibilities for collaboration between technology and creativity, allowing for a deeper exploration of the human mind's capacity for innovation and originality. By harnessing the power of neural signals, artists can create art that is truly a reflection of their innermost thoughts and emotions, showcasing the untapped potential of the human brain in the realm of artistic expression. The integration of BCI technology in the artistic process not only revolutionizes how art is created but also how it is experienced by audiences. By immersing viewers in interactive art installations controlled by neural signals, artists can create truly immersive and transformative experiences that blur the lines between creator and observer. This transformative potential of BCI technology in the art world highlights the symbiotic relationship between human creativity and technological innovation, opening up new avenues for exploration and expression that have the power to revolutionize the way we perceive and interact with art in the future.

## Ethical Implications of Neuroaesthetic Experiences

One of the ethical implications of neuroaesthetic experiences is the potential manipulation of individuals' sensory perceptions and emotions. Neuroaesthetics, which examines the neural basis of aesthetic experiences, raises questions about the authenticity and integrity of art encounters when technology is used to enhance or alter them. As advancements in neurotechnology enable the direct stimulation of brain regions associated with pleasure and beauty, individuals may be susceptible to engineered emotional responses that undermine the genuine appreciation of art. This challenges the ethical boundaries of sensory manipulation and the autonomy of individuals in determining their aesthetic preferences. The commodification of neuroaesthetic experiences poses ethical dilemmas regarding the commercialization of art and the exploitation of human neurobiology. As neurotechnology allows for individualized and amplified aesthetic experiences, there is a possibility for companies to capitalize on these enhanced sensations in the market. This raises concerns about the commercialization of art as a commodity that targets and manipulates consumers' neural responses for profit. The power dynamics between producers of neuroaesthetic content and consumers come into question, as the influence of technology on individuals' emotions and preferences may lead to the exploitation of vulnerable populations for economic gain. The privacy and consent issues surrounding neuroaesthetic experiences highlight the ethical complexities of accessing and manipulating individuals' neural data. As neurotechnology collects and analyzes brain activity to personalize aesthetic experiences, concerns arise about the protection and ownership of sensitive neural information. Privacy breaches and unauthorized access to neural data could have implications for individuals' autonomy and decision-making processes. Ensuring informed consent and data security in neuroaesthetic practices becomes essential to uphold ethical standards and respect individuals' rights to privacy and self-determination in the digital age. By addressing these ethical implications, society can navigate the evolving landscape of neuroaesthetics responsibly and ethically.

# XXIX. NEUROTECHNOLOGY AND NEUROETHICS IN ARTIFICIAL INTELLIGENCE (AI)

In the realm of neurotechnology and artificial intelligence, the integration of BCIs has opened up new possibilities for the interaction between humans and machines. The development of BCIs has a rich history that traces back to early research and fundamental discoveries in neuroscience. From rudimentary devices to advanced systems, the evolution of BCI technology has allowed for groundbreaking applications in various fields, ranging from medical treatments to entertainment and communication. As our understanding of the human brain continues to expand, so too does the potential of BCIs to enhance the way we interact with technology. One of the key aspects of BCI technology is its operation, which involves sensors, signal processing, and decoding algorithms to translate neural signals into actionable commands. The distinction between invasive and non-invasive BCIs also plays a critical role in determining the feasibility and accessibility of BCI applications. In the medical field, BCIs have shown promising results in the treatment of motor disabilities, neurorehabilitation for stroke patients, and potential interventions for neurological and psychiatric disorders. These advancements highlight the transformative power of BCIs in improving the quality of life for individuals affected by various health conditions. The integration of BCIs into everyday life raises important ethical and social challenges that must be addressed. Issues such as privacy and security of neural data, risks of technology dependence and abuse, and the ethical implications of direct manipulation of the human brain all require careful consideration. As we look towards the future of BCIs, it is crucial to navigate these challenges thoughtfully to ensure that the benefits of this technology are maximized while mitigating potential harms. By exploring the intersection of neurotechnology and neuroethics in artificial intelligence, we can pave the way for a more ethical and socially responsible integration of BCIs into our lives.

## Ethical Considerations in AI Integration with BCI Technology

Ethical considerations play a crucial role in the integration of AI with BCI technology. One of the primary concerns is the potential invasion of privacy through the collection and use of neural data. As BCI devices become more advanced, there is a risk that sensitive information about an individual's thoughts, emotions, and cognitive processes could be accessed without their consent. This raises questions about data security, ownership, and the ethical implications of using this data for commercial or research purposes. The ethical dilemmas extend to issues of autonomy and agency when AI is integrated with BCI technology. There is a concern that the use of AI algorithms to decode neural signals and make decisions on behalf of individuals could undermine their autonomy and free will. This blurring of boundaries between human consciousness and machine intelligence raises philosophical questions about what it means to be human and the ethical implications of merging our cognitive functions with artificial systems. It also raises concerns about the potential for manipulation and control by external actors who have access to the AI-BCI system. The integration of AI with BCI technology raises concerns about accountability and transparency in decision-making processes. As AI algorithms become more complex and opaque, it becomes challenging to understand how decisions are being made and who is ultimately responsible for the outcomes. This lack of transparency can lead to ethical issues related to bias, discrimination, and unintended consequences. It is essential for developers, researchers, and policymakers to carefully consider these ethical dimensions and ensure that AI-BCI systems are designed and used in a way that upholds the values of autonomy, privacy, and respect for individual rights.

## Implications of AI Algorithms on Neural Data Privacy

The implications of AI algorithms on neural data privacy are a key concern in the development and implementation of BCIs. As these algorithms become more sophisticated and capable of analyzing intricate neural patterns, the potential for unauthorized access to sensitive neural data increases. This raises significant privacy issues, as individuals may be at risk of having their private thoughts, emotions, and cognitive processes exposed without their consent. The use of AI algorithms in decoding neural signals for various applications, such as controlling external devices or predicting behavior, could lead to potential misuse of this information for commercial or even malicious purposes. In light of these privacy concerns, it is essential for researchers, developers, and policymakers to establish robust regulations and safeguards to protect neural data collected through BCIs. Encryption measures and secure data storage protocols must be implemented to prevent unauthorized access or hacking attempts. Transparency in the collection and use of neural data, as well as obtaining informed consent from users, are crucial ethical considerations in the development of BCI technologies. Addressing these implications of AI algorithms on neural data privacy is pivotal in ensuring that the benefits of BCIs can be maximized without compromising individual privacy and autonomy. Looking ahead, as AI continues to advance and become more integrated into BCI technologies, ongoing evaluation and adaptation of privacy measures will be necessary to safeguard neural data. Collaborative efforts from interdisciplinary teams, including experts in neuroscience, computer science, and ethics, are essential in navigating the complex landscape of neural data privacy. By proactively addressing these implications and developing ethical frameworks for the use of AI algorithms in BCIs, we can harness the transformative potential of neurotechnology while upholding the fundamental principles of privacy, autonomy, and data security.

## Ensuring Ethical AI Development and Implementation

One crucial aspect of advancing neurotechnology, particularly in the development and implementation of BCIs, is ensuring that ethical considerations are at the forefront of decision-making processes. As these technologies continue to evolve and become more integrated into daily life, it is imperative to address potential ethical dilemmas that may arise. One major concern is the issue of privacy and security related to neural data. With the ability to access and interpret brain activity, there is a significant risk of unauthorized access to sensitive information, leading to potential breaches of personal privacy. It is essential to establish robust ethical guidelines and secure data protection measures to mitigate these risks and safeguard users' privacy. The reliance on BCIs for communication, motor control, and other essential functions raises concerns about technology dependence and potential abuse. As individuals become more reliant on these devices for everyday tasks, there is a risk of losing autonomy and agency, as well as potential psychological effects from over-reliance on external technologies. Addressing these ethical challenges requires a careful balance between promoting innovation and technological development while also prioritizing the well-being and autonomy of individuals. By incorporating ethical considerations into the design, development, and implementation of BCIs, it is possible to mitigate potential risks and ensure that these technologies contribute positively to society. The ethical development and implementation of AI, including BCIs, are essential for harnessing the full potential of these technologies while minimizing potential risks and pitfalls. By prioritizing ethical considerations such as privacy, security, autonomy, and well-being, it is possible to create a framework that fosters innovation while upholding fundamental ethical principles. As BCIs continue to evolve and become more prevalent in various sectors, including healthcare, entertainment, and communication, it is crucial to have robust ethical guidelines in place to guide their responsible integration into society. By addressing ethical challenges proactively, we can ensure that neurotechnology enhances human capabilities, improves quality of life, and promotes positive societal change.

# XXX. NEUROTECHNOLOGY AND NEUROSECURITY

In today's rapidly advancing technological landscape, neurotechnology, specifically BCIs, has the potential to revolutionize the way humans interact with technology. By directly connecting the brain to external devices, BCIs hold promise for enhancing communication, control, and access to information for individuals with physical limitations. The development of BCIs has evolved from early research in neuroscience to sophisticated systems utilizing advanced sensors, signal processing, and decoding algorithms. This progression has paved the way for a range of applications in medical fields, such as the treatment of motor disabilities through mind-controlled prostheses, and neurorehabilitation for stroke patients. One of the key areas where BCIs are making significant strides is in the field of entertainment and communication. The ability to control video games and virtual reality experiences using only the power of the mind opens up new possibilities for immersive entertainment. BCIs are enabling enhanced communication for individuals with physical limitations, allowing them to express themselves more effectively. Looking ahead, the entertainment industry stands to benefit greatly from the integration of BCI technology, offering innovative and engaging experiences that were previously unimaginable. While the potential benefits of BCIs are vast, ethical and social challenges must also be considered as the technology continues to advance. Issues such as privacy and security concerning neural data, risks of technology dependence, and ethical implications of direct manipulation of the human brain need to be carefully addressed. By navigating these challenges thoughtfully, we can harness the transformative power of BCIs to improve quality of life and open up new possibilities for humanity in the future. As society engages with the ethical and societal implications of this cutting-edge technology, the co-evolution of humans and technology has the potential to shape a future where individuals can thrive in unprecedented ways.

## Cybersecurity Challenges in BCIs

One of the significant cybersecurity challenges in BCIs is the risk of unauthorized access to neural data. As BCIs require direct access to brain signals for operation, there is a concern regarding the privacy and security of this sensitive information. Hackers could potentially intercept and exploit neural data, leading to breaches of personal information and even manipulation of thoughts or actions. Ensuring robust encryption protocols and secure data storage methods are crucial in safeguarding against such cyber threats in BCI technology. The interconnected nature of BCIs with external devices and networks poses a vulnerability to cyber-attacks. As BCIs often communicate wirelessly with other devices or servers, there is a risk of interception or manipulation of data during transmission. Hackers could potentially hack into BCI systems to gain control over connected devices or inject malicious code, leading to harmful consequences. Implementing secure communication protocols and stringent access controls are essential in mitigating these cybersecurity risks and protecting the integrity of BCI operations. The convergence of AI with BCIs introduces additional cybersecurity challenges. AI algorithms are increasingly utilized in BCI systems for processing and interpreting neural signals, enhancing the overall performance and capabilities of the technology. AI-powered BCIs also present new avenues for cyber-attacks, such as adversarial attacks that can deceive AI algorithms or manipulate their output. Developing robust AI security mechanisms and conducting thorough vulnerability assessments are essential in ensuring the resilience of AI-integrated BCIs against emerging cyber threats. By addressing these cybersecurity challenges proactively, the potential benefits of BCIs can be maximized while safeguarding against malicious activities in this advanced neurotechnology domain.

## Safeguarding Neural Data from Unauthorized Access

In the realm of neurotechnology, safeguarding neural data from unauthorized access is a critical concern that must be addressed to ensure the ethical and secure use of BCIs. As these devices become more advanced and integrated into everyday life, the potential for malicious actors to exploit neural data for personal gain or harm increases. To mitigate this risk, robust encryption and cybersecurity measures must be implemented to protect the privacy and integrity of individuals' neural information. Strict regulations and guidelines should be established to govern the collection, storage, and dissemination of neural data to prevent unauthorized access and misuse. One approach to safeguarding neural data is the development of secure data transfer protocols and encryption algorithms specifically designed for BCIs. By encrypting neural signals at the source and ensuring that they can only be decoded by authorized parties, the risk of interception and unauthorized access can be significantly reduced. Implementing multi-factor authentication and access control mechanisms can further enhance the security of neural data, ensuring that only approved users can interact with the BCI system and access sensitive information. This layered approach to cybersecurity can help to protect neural data from unauthorized access and safeguard the privacy and confidentiality of individuals using BCIs. Safeguarding neural data from unauthorized access is paramount in the development and adoption of BCIs. By implementing robust encryption, cybersecurity measures, and strict regulatory frameworks, the privacy and integrity of individuals' neural information can be protected from malicious actors. As BCIs continue to advance and become more prevalent in various applications, addressing the security of neural data will be crucial to ensure the ethical and responsible use of this transformative technology. By prioritizing the protection of neural data, we can harness the full potential of BCIs while upholding the principles of privacy, security, and ethicality.

## Developing Secure Protocols for Neurotechnological Systems

As neurotechnological systems continue to advance, the importance of developing secure protocols for BCIs becomes paramount. These protocols are crucial for ensuring the privacy and integrity of neural data, as well as protecting users from potential security breaches. Implementing robust encryption techniques and authentication mechanisms can help safeguard sensitive information collected by BCIs, such as neural signals and brain activity patterns. By incorporating secure protocols into neurotechnological systems, researchers and developers can mitigate the risks associated with unauthorized access to neural data and maintain the trust of users in these cutting-edge technologies. Developing secure protocols for BCIs can also help address ethical concerns related to the direct manipulation of the human brain. Ensuring that neural data is encrypted and securely transmitted can prevent malicious actors from exploiting this information for unethical purposes, such as influencing a user's thoughts or actions without their consent. By establishing clear guidelines and best practices for securing neural data, the ethical implications of using BCIs for communication, entertainment, and medical applications can be carefully managed. This proactive approach to security can help promote responsible use of neurotechnological systems and protect the autonomy and agency of individuals using these devices. The development of secure protocols for neurotechnological systems is essential for unlocking the full potential of BCIs while safeguarding user privacy and security. By incorporating encryption, authentication, and data protection measures into BCIs, researchers and developers can enhance the reliability and trustworthiness of these innovative technologies. As society embraces the possibilities of interfacing directly with the human brain, it is imperative to prioritize the implementation of secure protocols to ensure that users can harness the benefits of BCIs without compromising their personal data or ethical values. By addressing these security challenges head-on, we can pave the way for a future where neurotechnology enhances human capabilities while upholding the principles of privacy and ethical integrity.

# XXXI. NEUROTECHNOLOGY AND NEURODIVERSITY ADVOCACY

Neurodiversity advocacy is a movement that seeks to promote the acceptance and inclusion of individuals with neurological differences such as autism, ADHD, dyslexia, and other neurodevelopmental conditions. The integration of neurotechnology, specifically BCIs, into the conversation around neurodiversity advocacy has the potential to revolutionize how these individuals interact with the world around them. By providing alternative methods of communication, control, and expression, BCIs can empower neurodiverse individuals to navigate daily tasks and engage with society on their own terms. This technological advancement not only enhances their quality of life but also challenges traditional notions of ability and disability, fostering a more inclusive and accommodating society. One of the key benefits of incorporating BCIs into neurodiversity advocacy is the ability to bridge communication barriers for nonverbal individuals with conditions such as severe autism. These individuals often face challenges in expressing their thoughts, feelings, and needs, leading to frustration and isolation. With the help of BCIs, they can now communicate through neural signals, allowing them to engage with others more effectively and have a voice in decision-making processes that directly impact their lives. This newfound ability to communicate can significantly improve their social connections, emotional well-being, and overall sense of autonomy and agency. The integration of BCIs in neurodiversity advocacy also highlights the importance of embracing and celebrating neurological differences as valuable contributions to society. Rather than viewing neurodiversity as a deficit or a limitation, the use of BCIs acknowledges the unique strengths and perspectives that neurodiverse individuals bring to the table. By championing the use of technology to support and amplify these strengths, neurodiversity advocacy can promote a more inclusive and equitable society that values diversity in all its forms. As we continue to explore the possibilities of neurotechnology in the context of neurodiversity advocacy, it is essential to prioritize accessibility, empowerment, and respect for individual differences to truly realize the transformative potential of BCIs in enhancing the lives of neurodiverse individuals.

## Promoting Inclusivity and Accessibility for Neurodiverse Individuals

Within the realm of neurotechnology and BCIs, promoting inclusivity and accessibility for neurodiverse individuals is paramount. Neurodiversity encompasses a range of neurological differences, including autism, ADHD, and dyslexia, among others. By fostering environments that cater to the diverse needs of neurodiverse individuals, we can create more inclusive spaces that allow for equal participation and opportunities. This involves accommodating sensory sensitivities, providing clear communication methods, and offering support tailored to individual requirements. Through the implementation of inclusive practices, we can ensure that neurodiverse individuals are not only included but valued for their unique perspectives and contributions to society. One key aspect of promoting inclusivity for neurodiverse individuals within the realm of neurotechnology is through the design of user-friendly interfaces that cater to a wide range of cognitive abilities. This involves incorporating features such as customizable settings, straightforward navigation options, and clear visual cues to enhance accessibility. By prioritizing user experience and usability in the development of neurotechnological devices, we can minimize barriers to participation for neurodiverse individuals and empower them to engage with technology on their own terms. Providing comprehensive training and support resources can further facilitate the integration of neurodiverse individuals into the digital landscape, ensuring that they can fully leverage the benefits of neurotechnological advancements. Advocating for inclusivity and accessibility for neurodiverse individuals in the context of neurotechnology is essential for creating a more equitable and inclusive society. By recognizing and addressing the unique needs and challenges faced by neurodiverse individuals, we can pave the way for greater participation, engagement, and empowerment within the realm of BCIs and other neurotechnological innovations. Through collaborative efforts and a commitment to universal design principles, we can build a future where neurodiversity is embraced and celebrated, leading to a more diverse and inclusive technological landscape for all.

## Advocating for Ethical Use of BCI Technology in Neurodivergent Communities

Neurodivergent communities, including individuals with conditions such as autism, ADHD, and dyslexia, stand to benefit greatly from advancements in BCI technology. By advocating for the ethical use of this technology within these communities, we can empower neurodivergent individuals to enhance their communication, independence, and overall quality of life. BCIs can provide alternate means of communication for non-verbal individuals, allowing them to express their thoughts and feelings more effectively. This can help bridge the communication gap often faced by those with neurodevelopmental disorders, promoting inclusivity and understanding. The ethical use of BCI technology in neurodivergent communities can also lead to improved access to education and employment opportunities. By leveraging BCIs to support cognitive functions such as attention, memory, and learning, individuals with neurodivergent conditions can overcome some of the challenges they may face in traditional academic and professional settings. BCIs can assist with focus and concentration, which are areas of difficulty for many individuals with ADHD. This can potentially level the playing field and create a more inclusive environment for neurodivergent individuals to thrive. Advocating for the ethical use of BCI technology in neurodivergent communities is not only a matter of social justice and inclusion but also a means of unlocking the full potential of individuals with neurodevelopmental conditions. By ensuring that the development and deployment of BCIs are guided by ethical considerations and respect for individual autonomy, we can harness the transformative power of this technology to enhance the lives of neurodivergent individuals. Through continued research, collaboration, and advocacy efforts, we can pave the way for a future where BCI technology is truly accessible and beneficial for all members of society, regardless of their cognitive differences.

## Addressing Stigma and Discrimination through Neurodiversity Awareness

Neurodiversity awareness plays a crucial role in addressing the stigma and discrimination often faced by individuals with neurological differences. By promoting the understanding that neurodivergent individuals have unique strengths and perspectives, society can embrace diversity and foster inclusivity. This shift in mindset can lead to greater acceptance and support for those with conditions such as autism, ADHD, and dyslexia, ultimately helping to break down barriers and create more inclusive environments. Educational initiatives and advocacy efforts focused on neurodiversity can help challenge stereotypes and promote a culture of acceptance and respect. Increasing awareness about neurodiversity can lead to the development of more tailored support services and accommodations for individuals with neurological differences. By recognizing the diverse needs and abilities of neurodivergent individuals, organizations can implement inclusive practices that ensure equal opportunities for success. This may involve providing specialized training, creating sensory-friendly environments, or offering alternative communication methods. Through these efforts, society can work towards creating a more equitable and accessible world for individuals with neurodevelopmental conditions, promoting their well-being and enhancing their quality of life. Embracing neurodiversity can foster a sense of belonging and community for individuals who may have previously felt marginalized or isolated. By celebrating the unique talents and perspectives of neurodivergent individuals, society can benefit from a rich tapestry of skills and experiences. Encouraging inclusion and fostering a sense of acceptance can empower individuals with neurological differences to thrive and contribute meaningfully to their communities. By promoting neurodiversity awareness, society can pave the way for a more inclusive and compassionate future where all individuals are valued and respected for their unique abilities and contributions.

# XXXII. CONCLUSION

The development and advancement of BCIs hold significant transformative potential for humanity. As discussed throughout this essay, BCIs have made remarkable progress in the fields of medical treatment, entertainment, communication, and beyond. The ability to directly interact with technology using only the power of the mind opens up new possibilities for individuals with physical limitations and neurological disorders. The integration of BCIs into everyday life has the potential to revolutionize how we interact with the world around us, providing a glimpse into a future where technology seamlessly enhances our abilities and quality of life. It is crucial to address the ethical and societal challenges that come with the widespread adoption of BCIs. Privacy concerns, technology dependence, and the ethical implications of directly manipulating the human brain are all issues that must be carefully considered and regulated. As the technology continues to advance, it is imperative that stakeholders work together to establish guidelines and regulations that protect individuals' rights and ensure the responsible use of BCIs. By tackling these challenges head-on, we can harness the full potential of BCIs while mitigating potential risks and harm. In spite of these challenges, the future of BCIs is bright. As emerging trends and technological advances continue to shape the landscape of neurotechnology, the co-evolution of humans and technology presents a promising future. By embracing the opportunities that BCIs offer, we can not only improve the quality of life for individuals but also explore new frontiers in human-computer interaction. The optimistic view is that BCIs will open up a world of possibilities, enhancing our capabilities and ultimately reshaping the way we live, work, and communicate. The journey towards this future is filled with exciting possibilities, and it is essential that we navigate it with mindfulness and collaboration.

## Summary of Key Findings on Neurotechnology and BCIs

Neurotechnology and BCIs have undergone significant advancements in recent years, leading to groundbreaking discoveries and innovative applications. One key finding is the successful use of BCIs in the treatment of motor disabilities, such as mind-controlled prostheses for amputees. These devices have enabled individuals with physical limitations to regain mobility and independence, marking a major breakthrough in the field of neurorehabilitation. BCIs hold promise for the treatment of neurological and psychiatric disorders, offering a non-invasive and personalized approach to managing conditions such as epilepsy and depression. The integration of BCI technology in the entertainment industry has opened up new possibilities for immersive experiences and enhanced communication. From mind-controlled video games to virtual reality simulations, BCIs are transforming how we interact with entertainment media. In the field of communication, BCIs have facilitated improved interfaces for individuals with speech and motor impairments, providing them with a means to express themselves more effectively. As the technology continues to evolve, the potential for BCIs to revolutionize entertainment and communication remains a compelling area of research and development. Alongside these exciting advancements, ethical and social challenges must be addressed to ensure the responsible and beneficial use of neurotechnology and BCIs. Privacy concerns related to the collection and storage of neural data, as well as the potential risks of technology dependence and manipulation, are critical issues that require careful consideration. The ethical implications of directly interfacing with the human brain raise important questions about autonomy, consent, and the implications of altering cognitive processes. By navigating these challenges thoughtfully and ethically, the future of BCIs holds tremendous promise for enhancing human capabilities and transforming the way we interact with technology.

## Reflection on the Transformative Potential and Ethical Considerations

As we delve deeper into the realm of BCIs, it becomes evident that the transformative potential of this technology is immense. The ability to directly interface with the brain opens up a world of possibilities, from restoring motor functions for individuals with disabilities to enhancing communication and entertainment experiences. With great power comes great responsibility, and ethical considerations must be front and center in the development and implementation of BCIs. As we navigate this new frontier, it is crucial to carefully consider the implications of direct manipulation of the human brain and the potential risks and consequences that come with it. One of the key ethical considerations when it comes to BCIs is the issue of privacy and security of neural data. With the ability to access and interpret signals directly from the brain, there is the potential for sensitive information to be exposed or misused. Safeguarding the privacy of individuals' neural data must be a top priority to prevent unauthorized access or exploitation. The risks of technology dependence and abuse must be carefully monitored and mitigated to avoid potential harm or manipulation of individuals through their neural interfaces. Despite the ethical challenges that come with the development of BCIs, there is a sense of optimism about the future of this technology and its potential to improve the quality of life for many individuals. By addressing ethical considerations head-on and proactively designing safeguards and regulations, we can harness the transformative power of BCIs while minimizing the risks. The co-evolution of humans and technology through BCIs offers a glimpse into a future where new possibilities for communication, entertainment, and medical treatments can enhance our lives in ways we have yet to fully imagine.

## Call to Action for Continued Research and Ethical Reflection in Advancing Neurotechnological Innovations

As neurotechnology continues to advance and shape the future of human-machine interaction, there is an urgent call to action for continued research and ethical reflection in the field of BCIs. While the potential benefits of BCI technology are vast, including medical applications in neurorehabilitation and the treatment of neurological disorders, as well as entertainment and communication enhancements, it is crucial to address the ethical and social challenges that come with such innovations. Privacy concerns, risks of technology dependence, and the ethical implications of direct manipulation of the human brain must be carefully considered and navigated through ongoing research and reflection. As BCI technology evolves and becomes more integrated into everyday life, it is essential to anticipate and prepare for emerging trends and technological advances. The co-evolution of humans and technology raises questions about the impact of BCIs on societal norms, individual autonomy, and overall quality of life. By staying ahead of these developments through proactive research and ethical consideration, we can ensure that the transformative potential of BCI technology is harnessed responsibly and ethically for the benefit of humanity. The future of BCIs holds great promise for enhancing human capabilities and experiences, but it also presents complex ethical and societal challenges that must be addressed. By continuing to push the boundaries of research in neurotechnology and engaging in ongoing ethical reflection, we can navigate these challenges thoughtfully and pave the way for a future where BCI technology improves lives and opens up new possibilities for humanity. It is crucial that researchers, policymakers, and society as a whole work together to ensure that the integration of BCIs into everyday life is done in a way that upholds ethical principles and advances the well-being of individuals and society as a whole.

# BIBLIOGRAPHY

Raman K. Attri. 'The Models of Skill Acquisition and Expertise Development.' A Quick Reference of Summaries, Speed To Proficiency Research: S2Pro©, 3/30/2019

Claudia Voelcker-Rehage. 'Cognitive and Brain Plasticity Induced by Physical Exercise, Cognitive Training, Video Games and Combined Interventions.' Soledad Ballesteros, Frontiers Media SA, 7/5/2018

Jane Hampton. 'Neuroplasticity.' Brain Training and Neuroscience Truths, Self Publisher, 11/14/2019

William H. Boothby. 'New Technologies and the Law in War and Peace.' Cambridge University Press, 1/1/2019

Laurel Allender. 'Designing Soldier Systems.' Current Issues in Human Factors, John Martin, CRC Press, 5/20/2018

Ian Ritchie. 'Neuroarchitecture.' Designing with the Mind in Mind, John Wiley & Sons, 12/21/2020

Thomas D. Parsons. 'Virtual and Augmented Reality methods in Neuroscience and Neuropathology.' Valerio Rizzo, Frontiers Media SA, 12/30/2020

Dario Robleto. 'Mobile Brain-Body Imaging and the Neuroscience of Art, Innovation and Creativity.' Jose L. Contreras-Vidal, Springer Nature, 11/15/2019

Oshin Vartanian. 'Neuroaesthetics.' Martin Skov, Routledge, 2/6/2018

Cynda Hylton Rushton. 'Moral Resilience.' Transforming Moral Suffering in Healthcare, Oxford University Press, 10/2/2018

United States. Congress. House. Committee on Merchant Marine and Fisheries. Subcommittee on Fisheries and Wildlife Conservation and the Environment. 'Growth and Its Implications for the Future.' Hearing with Appendix, Ninety-third Congress, First[-second] Session...., U.S. Government Printing Office, 1/1/1973

Jaime Wood. 'The Word on College Reading and Writing.' Carol Burnell, Open Oregon Educational Resources, 1/1/2020

Division of Neuroscience and Behavioral Health. 'Bridging Disciplines in the Brain, Behavioral, and Clinical Sciences.' Institute of Medicine, National Academies Press, 9/24/2000

Joseph M. Kizza. 'Ethical and Social Issues in the Information Age.' Springer Science & Business Media, 3/9/2013

Lisa Rosner. 'The Technological Fix.' How People Use Technology to Create and Solve Problems, Routledge, 2/1/2013

Sharmin Hossain. 'Neuroethics.' Anticipating the Future, Judy Illes, Oxford University Press, 1/1/2017

Owen J. Flanagan. 'Neuroexistentialism.' Meaning, Morals, and Purpose in the Age of Neuroscience, Gregg D. Caruso, Oxford University Press, 1/1/2018

Adrienne Colella. 'Neurodiversity in the Workplace.' Interests, Issues, and Opportunities, Susanne M. Bruyère, Taylor & Francis, 7/1/2022

James Giordano. 'Neurotechnology in National Security and Defense.' Practical Considerations, Neuroethical Concerns, CRC Press, 9/25/2014

Maya Bialik. 'Artificial Intelligence in Education.' Promises and Implications for Teaching and Learning, Wayne Holmes, Center for Curriculum Redesign, 1/1/2019

Leigh Richardson. 'Turn Your Brain on to Get Your Game On.' The How, What, Why to Peak Performance, Clovercroft Publishing, 1/7/2020

Wolfgang Broll. 'Virtual and Augmented Reality (VR/AR).' Foundations and Methods of Extended Realities (XR), Ralf Doerner, Springer Nature, 1/12/2022

Anton Nijholt. 'BCIs.' Applying our Minds to Human-Computer Interaction, Desney S. Tan, Springer Science & Business Media, 6/10/2010

*Veljko Dubljevic. 'Cognitive Enhancement.' Ethical and Policy Implications in International Perspectives, Fabrice Jotterand, Oxford University Press, 5/9/2016*

*Jeroen J. G. van Merriënboer. 'Training Complex Cognitive Skills.' A Four-Component Instructional Design Model for Technical Training, Educational Technology, 1/1/1997*

*Mikhail A. Lebedev. 'Modern Approaches to Augmentation of Brain Function.' Ioan Opris, Springer Nature, 8/25/2021*

*Allen Coin. 'Policy, Identity, and Neurotechnology.' The Neuroethics of BCIs, Veljko Dubljević, Springer Nature, 4/26/2023*

*Stephen Wear. 'Informed Consent.' Patient Autonomy and Clinician Beneficence Within Health Care, Georgetown University Press, 1/1/1998*

*Kristen J. Mathews. 'Proskauer on Privacy.' A Guide to Privacy and Data Security Law in the Information Age, Practising Law Institute, 1/7/2017*

*Paul C. Lebby. 'Brain Imaging.' A Guide for Clinicians, OUP USA, 4/1/2013*

*Felix Aplin. 'Prostheses for the Brain.' Introduction to Neuroprosthetics, Andrej Kral, Elsevier Science, 4/9/2021*

*Nick Ward. 'Oxford Textbook of Neurorehabilitation.' Volker Dietz, Oxford University Press, 1/1/2015*

*Alkinoos Athanasiou. 'Neurotechnology.' Methods, advances and applications, Victor Hugo C. de Albuquerque, Institution of Engineering and Technology, 4/26/2020*

*Reza Fazel-Rezai. 'BCI Systems - Recent Progress and Future Prospects.' IntechOpen, 1/1/2013*

*Brendan Z. Allison. 'BCIs.' Revolutionizing Human-Computer Interaction, Bernhard Graimann, Springer Science & Business Media, 10/29/2010*

*Rajesh P. N. Rao. 'Brain-Computer Interfacing.' Cambridge University Press, 9/30/2013*

Gernot Müller-Putz. 'BCI Research.' A State-of-the-Art Summary 4, Christoph Guger, Springer, 12/12/2015

Anton Nijholt. 'Brain–Computer Interfaces Handbook.' Technological and Theoretical Advances, Chang S. Nam, CRC Press, 1/9/2018

Institute of Medicine. 'From Neurons to Neighborhoods.' The Science of Early Childhood Development, National Research Council, National Academies Press, 11/13/2000

P. M. S. Hacker. 'History of Cognitive Neuroscience.' M. R. Bennett, John Wiley & Sons, 8/15/2012

NARAYAN CHANGDER. 'RESEARCH METHODOLOGY.' THE AMAZING QUIZ BOOK, Changder Outline, 12/21/2022

Ahmad Taher Azar. 'BCIs.' Current Trends and Applications, Aboul Ella Hassanien, Springer, 11/1/2014

James Giordano. 'Neurotechnology.' Premises, Potential, and Problems, CRC Press, 4/26/2012

www.ingramcontent.com/pod-product-compliance
Lightning Source LLC
Chambersburg PA
CBHW071829210526
45479CB00001B/51